THE PACIFIC RIM IN THE 21ST CENTURY

Alliances and Collaborations in the Asia-Pacific

THE PACIFIC RIM IN THE 21ST CENTURY

Alliances and Collaborations in the Asia-Pacific

With an introduction by

Dr. Mahathir Mohamad

Prime Minister of Malaysia

Pelanduk Publications

ASLI
ASIAN STRATEGY & LEADERSHIP INSTITUTE

Published by
Asian Strategy & Leadership Institute,
Level 1, Menara Sungei Way, Jalan Lagun Timur,
Bandar Sunway, 46150 Petaling Jaya,
Selangor Darul Ehsan, Malaysia.

With
Pelanduk Publications (M) Sdn. Bhd.,
24 Jalan 20/16A, 46300 Petaling Jaya,
Selangor Darul Ehsan, Malaysia.

Address all correspondence to
Pelanduk Publications (M) Sdn. Bhd.,
P.O. Box 8265, 46785 Kelana Jaya,
Selangor Darul Ehsan, Malaysia.

This book is a compilation of a selection of papers presented at the Pacific Rim Business Collaboration Symposium held in Kuala Lumpur, Malaysia on December 5-8, 1993.

Perpustakaan Negara Malaysia   Cataloguing-in-Publication Data

The Pacific Rim in the 21st century: alliances and collaborations
in the Asia-Pacific / Asian Strategy & Leadership Institute;
with an introduction by Mahathir Mohamad.
ISBN 967-978-433-9
1. Economic forecasting—Pacific Area. 2. Pacific Area—
Economic policy—2000- . 3. Asia—Foreign economic relations.
4. Pacific Area—Foreign economic relations. 5. Investments,
Foreign—Asia. 6. Asia—Economic integration. I. Pacific Rim
Business Collaboration Symposium (1993: Kuala Lumpur,
Malaysia). II. Asian Strategy & Leadership Institute.
III. Mahathir Mohamad, Dato' Seri, 1926- .
337.11823

Printed in Malaysia by
Academe Art & Printing Services Sdn. Bhd.

Contents

Preface
The Pacific Rim in the 21st Century

MIRZAN MAHATHIR

WITH NAFTA set to extend from Canada to Chile and as more and more nations qualifying as Asian Tigers, the world's economic hub is moving slowly but surely to the Pacific Rim. Many experts and strategists have predicted that the Asian Century is about to begin. At the same time, the initial statistics following NAFTA's implementation are just coming in and they indicate substantial increase in trade.

The trends are there for all to see. We are witnessing increasing economic interdependence, the growing needs of developing countries and the disintegration of barriers to the flow of capital, information and technology across borders. These are the factors fuelling the drive towards global economic and market integration. Organizations need to refine their corporate and marketing strategies, reposition themselves and offer new value-added products and tailor-made services to capitalize on new business opportunities globally.

The Pacific Rim continues to be the major engine of growth for the global economy. Two-thirds of the world's wealth is concentrated in this region. The Asia-Pacific region is a US$3-trillion market with 1.8 billion potential customers. The opportunities for collaboration among parties from the Asean member-nations, China, Korea, Hong Kong and Japan on the western shores of the

Pacific and Canada, Mexico, the United States and Chile on its eastern shores are immense.

Recognizing the growing importance of the Pacific Rim and its emergence as the major driving force in the global economy, the Asian Strategy & Leadership Institute (ASLI) in collaboration with International Infotrade, Inc., a Hawaii-based consulting group, organized the first-ever Pacific Rim Business Collaboration Symposium in Kuala Lumpur in December 1993. The symposium was intended to promote strategic alliances and business collaborations among countries of the Pacific Rim. It brought together cabinet ministers, captains of industry, entrepreneurs, investors, senior government officials, leading academicians and top consultants to discuss the challenges, prospects and strategic issues facing this region as we approach the twenty-first century.

The Prime Minister of Malaysia, Dato' Seri Dr Mahathir Mohamad, delivered the keynote address at the symposium. Among other distinguished personalities who addressed the symposium were Malaysia's International Trade and Industry Minister, Dato' Seri Rafidah Aziz, Singapore's Information and Arts Minister, Brigadier-General George Yeo, Australia's Trade Minister, Senator Peter Cook, and the Executive Vice-President for Asia of the Mexican Investment Board, Jose Landa. The symposium was significant in that it emphasized business collaborations and strategic alliances among nations in the region.

The publishing of this book is a joint effort between ASLI and Pelanduk Publications. It contains a selection of papers presented at the symposium and other relevant topics that could serve as a useful guide for entrepreneurs, investors, business managers and those interested in forging business linkages in the dynamic Pacific Rim. It is my hope that companies and businessmen in this region will build on ties established and work to strengthen cross-border strategic alliances. Through such collaborations, profitable opportunities can be developed, goodwill generated and suspicions reduced, thereby contributing to regional stability and long-term security.

Indeed, the high-growth Asean nations, the Asian Tigers, the significant developments in China, Mexico and Chile, and the sophisticated markets of Japan, the United States and Canada, to name a few, all point towards the growing need for effective presence. The ever-increasing investment and trade flows within the region mean that more governments and companies are positioning themselves to reap the benefits of doing business in the Pacific Rim.

The co-operation and support of Pelanduk Publications were invaluable. We also wish to thank the Prime Minister of Malaysia, Dato' Seri Dr Mahathir Mohamad, for delivering the opening and keynote address at the symposium. The full text of Dr Mahathir's address has been incorporated as an introduction to this book. Our sincere appreciation also goes to all speakers and panellists who contributed to the success of the symposium. I also wish to acknowledge the support of ASLI's Special Projects staff in putting together both the symposium and this book.

This book serves as a testimony to the efforts of the many who contributed to the symposium and will be a useful guide for anyone interested in doing business in the Pacific Rim. I hope that this book will present some useful ideas and add to the reader's knowledge of the key challenges, strategic issues and opportunities in this part of the world.

Mirzan Mahathir,
Executive Chairman,
Asian Strategy & Leadership Institute,
Kuala Lumpur, Malaysia.

Introduction

Alliances and Collaborations within the Pacific Rim

MAHATHIR MOHAMAD

THE hub of the world's economic activity is gravitating towards the Pacific Rim. *The Pacific Rim in the 21st Century: Alliances and Collaborations in the Asia-Pacific* looks at how the region can make its mark as an increasingly powerful force in the global economy through intra-regional strategic alliances and collaborations such as joint ventures, business partnerships and technology transfers. Such global business partnerships are changing the way business is conducted around the world. And as the twentieth century draws to a close, the Pacific Rim will be the major engine of growth powering the world economy, based on current performances. This is especially so because of the increasing size of the Pacific Rim market. Two-thirds of the world's wealth is in the Pacific Rim: a US$3-trillion market with 1.8 billion potential customers. The dynamism and tremendous opportunities available in the region should pave the way for investors to acquire a competitive edge.

It is also a fact that despite poor economic performances elsewhere, East Asian nations have continued experiencing high growth and it would not be wrong to anticipate that this bullish trend will continue for many more years to come. Indeed, it was not too long ago when we were told that the Mediterranean was the ocean of the past; the Atlantic, the ocean of the present; and the Pacific, the ocean of the future. The future is here now and Asia's share of global gross domestic product (GDP) is expected to increase from about 25 per cent to almost 33 per cent by the twenty-first century. Asia's economic growth has clearly outstripped that of the world.

Malaysia has been able to transform itself from an agricultural-based, commodity-driven economy into a diversified manufacturing-based, export-led economy. Today, Malaysia is one of the few countries in the world that enjoy buoyant economic growth with low inflation. While it is true that some countries are booming they are also afflicted with runaway inflation. In Malaysia, prudent economic and fiscal policies have helped businesses to prosper and grow rapidly without affecting the low rate of inflation characteristic of the country. With our transparent and pragmatic policies, we hope to encourage greater inflow of foreign investments involving sophisticated technology, high capital, technical know-how and marketing expertise. The private sector should further enhance its capacity to tap the potential markets of the Asian countries while establishing a symbiotic and mutually beneficial relationship.

Today, Malaysia is modestly proud to be regarded a model of economic development. If I may say so the achievement is even more impressive as we have managed not only to promote rapid economic development, but have done so despite being a nation of diverse and incompatible ethnic and religious mix.

Malaysia's economic transformation and East Asia's success story did not come about by accident. It was also not sheer luck that has made us what we are today. Our success story is the product of much hard work, together with a work ethic that is compatible with industrial peace and high productivity.

Indeed we believe a strong, stable government is a major prerequisite for success. Look around you in East Asia and you will find that there is a strong correlation between rapid economic growth and strong stable governments. Strong governments are prepared to take unpopular decisions in the best interest of the nation. Strong, stable governments adopt a longer-term outlook on macro-planning and are not just concerned with the next general elections. Strong governments provide the necessary stability and predictability so essential to long-term investments.

Strong governments do not mean the absence of free, democratic elections. When people understand the limitations and the responsibility of citizens, democracy can be practised without democratic extremism and anarchy. Democracy is meant to serve the people, not the people to serve democracy. When the devotion to democracy results in a stagnant economy, high unemployment and denial of the right to work and work hard; when democracy protects fascists and neo-Nazis; when the individual activist takes precedence over the silent masses then it is time to question whether we have correctly interpreted democracy. It is important to remember that fanatical democrats are no better than religious fanatics for they both cannot see the wood for the trees.

In the West, democracy means many things to different people. To Asians, democracy means our citizens are entitled to free and fair elections. They can choose the governments of their liking. We also believe that once we have elected our governments they should be allowed to govern and to formulate policies and act on them. Our democracy does not confer complete licence for citizens to go wild. We need political stability, predictability and consistency to provide the necessary environment for progress and economic development. Thus, we have seen the Singapore story unfolded with Lee Kuan Yew at its helm and now continued by his successor, Goh Chok Tong. We see this in Indonesia's success story with President Suharto staying at the helm for almost thirty years.

Malaysia is socially stable. Although our people differ in terms of culture, religion and philosophy, they display tolerance and understanding and respect each others' way of life. Malaysians generally believe in compromising. We do not compromise out of weakness but out of a common desire to see our nation remain peaceful and prosperous. The vast majority of Malaysians are moderate, tolerant, sensitive to each other's differences and harbour little ill will.

Malaysia is fortunate to have a resilient private sector willing to undertake risks and share the tasks of building economic prosperity. The private sector is the main engine of growth for the Malaysian economy. With the government's proactive and sound macroeconomic principles and programmes, the concept of Malaysia Incorporated and privatisation, we have been able to accelerate growth, reduce foreign debts, deregulate and assist the private sector to expand their role. We have invested in the future by allocating vast sums of money to build infrastructure, as well as provide education and training suitable for an industrialized economy. It is this commitment towards investing in infrastructure and human resource development that has given us a strong competitive edge.

Malaysia will continue to maintain its liberal economic policies and development programmes. Our policies will remain consistent, predictable and transparent. For us, the challenge of managing success is just as important as managing problems during a recession. I am confident that with the positive measures we have taken, we are today very much better equipped to cope with any cyclical downturn.

However, to enable us to better manage difficult times if a downturn occurs, we must not fritter away our gains during boom times. We will continue to increase national savings and diversify our economy so that we will not be overdependent on any one sector. We must diversify our markets as well as seek new markets for our products and services. In fact, today 53 per cent of Malaysia's trade is with the nations of East Asia where once it was almost exclusively with Great Britain.

Rapid growth usually leads to overheating. We must not allow interest rates to be too high that it might stifle investments. We must not allow wages to outpace productivity. High inflation, breakdown in services, inadequate funds for needed supporting infrastructure, overstrained bureaucracy and inadequate or shortage in the supply of trained personnel; all these alone or together can retard or even reverse economic growth. How a government deals with overheating will determine the future of the nation and its own fate.

The trend today is towards a free market economy. China, the republics of the former Soviet Union, the Indo-China states and the ex-Communist countries of Eastern Europe have all embraced the free market system. Some have even adopted the democratic system of government. The mere adoption of these economic and political systems is no guarantee of success. If they are, then all the western democracies with market economies would be eternally prosperous. But we all know that they are not. In fact, the countries which are doing well now are those with not so liberal a democracy and an economic system in which the governments play a significant role. China and Vietnam, which look set to take off economically, have approximately this kind of combination. It would be unfortunate if in their zeal to proselytize, the victorious democrats inflict economic and political disaster on the enthusiastic converts. It must be warned that democracy and free markets are not economic and political cure-alls.

Although, East Asia has achieved remarkable success and has demonstrated strong, resilient growth, the future is full of uncertainties even for them. Calamities and disasters are always possible even for those with brilliant records. Japan is the contemporary example. It is therefore even more crucial and imperative that East Asian nations work together in a more integrated and cohesive manner. The concept of an East Asia Economic Caucus (EAEC) must be strengthened further. As I have reiterated before, the EAEC is GATT-consistent and is committed to free trade. The EAEC is a result of and intended for open regionalism. On one hand, it recognizes the need for regional co-operation and integration and, on the other, it promotes free trade. East Asian countries must work together. The need to balance regional groupings elsewhere is obvious. Alone, no Asian country, however powerful, can match the combined clout of the economic groupings which have been formed in Europe and North America.

No one need fear an Asian grouping. Europeans and North Americans are much more homogeneous than Asians. Even in East Asia, we are not only ethnically different but we are also divided by culture, language and religion. It is quite impossible for Asians to think of themselves the way Europeans think of themselves as being one race and one colour. Indeed the European culture and language have common origins. So the idea of an integrated Asian community borders on the ridiculous. Not so an Asian or an East Asian caucus designed to discuss common economic problems and to a limited extent to help pull up those among the East Asian countries which are lagging behind. All these can be motivated by enlightened self-interest. If Southeast Asia is one of Japan's valuable markets today, it is because Japan invested in these countries initially and helped them to develop. The successful economies of East Asia will be helping themselves when they help the less developed countries in the region to grow and prosper.

The EAEC is therefore a logical follow-up to the economic interaction in East Asia in the years after the Second World War. The EAEC is good for the governments of East Asia which are largely concerned to see that their countries achieve economic well-being. The EAEC is of course good for the business community of East Asia as well. What else do businessmen ask for more than to spread their wings beyond the borders of their countries.

Today, the narrow protectionist outlook of Asian businessmen has all but disappeared. Not only are they keen to go abroad but they regard business alliances and collaborations with companies in other countries as good business strategy. By doing so they can gain access to new markets and also acquire new technology. Indeed, there is much to be gained through cross-border economic alliances and collaborations regionally and internationally.

1

Forging Successful Strategic Business Alliances

MICHAEL YEOH

TOMORROW'S business competition will be won by companies that thrive on collaborations and strategic alliances. Increasingly, we see more consortiums being established, more businessmen working together, because such partnerships are the way to succeed in the global marketplace.

Successful business partnerships or strategic collaborations depend on relationships. Establishing and maintaining a good relationship is the key to successful strategic partnerships. This will require parties to collaborate in win-win ventures where everyone will stand to gain or benefit. The benefits to be gained should not be loaded in favour of only one party, with the other party gaining much less.

In this regard, I am reminded of the story of the two boys who were strolling in a park when they encountered a bear. One of them suggests running away while the other quietly takes his running shoes out from his backpack and starts to put them on. The first boy started saying, "Don't be silly. You can't possibly outrun the bear." The second boy looked up at his partner and replies, "I don't have to outrun the bear—all I need to do is to run faster than you."

This story illustrates a lopsided partnership, where two good friends in the face of imminent danger do not work together to overcome the problem, but went separate ways to compete and survive. Very often, as Edward de Bono, the lateral thinker points out, there is danger in competitive thinking when actions and strategies are determined by the behaviour of competitors.

Strategic alliances will grow in importance because of the vital need for companies to search for new markets globally and the increasing tempo of global competition. Furthermore, there is a need to innovate and to come out with new products and services so as to reduce product development time.

Strategic alliances are used for a variety of compelling reasons: to reduce costs in product development and new business opportunities, to reduce risks in expanding into new markets and new businesses, to enter and penetrate markets faster through collaborative research and product innovation, to build and maintain flexibility, and to neutralize competition by collaborating instead of competing with competitors. Successful partnering depends on both sides agreeing to the same approach and having a common vision as well as common values. Establishing mutually agreeable goals are fundamental to seeking a common vision. Often business partnerships flounder because of poor direction and a lack of clear goals. Setting goals that are mutually agreeable is vital.

Goal-setting involves clarifying expected outcomes, determining what the results expected are and understanding who should do what to make the partnership successful. Successful partnerships occur when the partners' strategic goals converge.

Some fundamental questions to consider that can facilitate goal-setting are: what are the most critical business goals or corporate objectives to be achieved from the partnership or alliance? what are the current common problems or areas of concern? what do we need to improve on? what can I contribute to the alliance? how can the other party benefit from my contribution? what must I do to make the alliance successful? what are our action plans to carry out the alliance? and how do we resolve conflicts and where do we go when we have problems?

Some attributes for successful strategic alliances include: building trust and mutual confidence, be open or maintain ethical communication, be ethical and honest as integrity is important, think long-term—do not only look for short-term benefits, and be a team player—it takes everybody to win.

We have identified several key factors for successful business partnerships: top-level commitment from both sides, willingness to compromise and seek mutual understanding, clearly understood grievance-handling procedures, and both partners must recognize that they can learn from each other. More important is the mutual desire to succeed.

Strategic alliances can take many forms. Some alliances are *ad-hoc* arrangements while others are project-based joint ventures. Consortiums can be another form of strategic alliance. Usually, a company that gets involved in an alliance is keen to take advantage of economies of scale, to gain access to knowledge or technology or to perform better.

Strategic alliances should be encouraged in Malaysia. Foreign multinational corporations can forge relationships in Malaysia to facilitate the transfer of technology. Alliances between *bumiputera* and non-*bumiputera* companies can further strengthen national unity through genuine and sustained multiracial economic co-operation.

The Malaysia Inc. concept can also be likened to a strategic alliance between the public and private sectors. The sharing of ideas, exchange of information and pooling of experiences can enable the Malaysia Incorporated exercise to be an effective strategic alliance, contributing to national growth and development.

George Yip has identified several key strategies to successfully manage strategic alliances. First, you must have a clear understanding of the company's current capabilities and the capabilities it will need in the future. Second, you must also consider a wide range of possible alliances. Third, before committing the company to an alliance, scrutinise the values, commitment and capabilities of prospective partners. Fourth, understand the risks of opportunism, knowledge-leaks and obsolescence. Fifth, structure and manage alliances as separate entities. Sixth, partners must come to trust each other. And last, but not least, alliances must be led, not managed.

2

Southeast Asia in the Pacific Rim Century

GEORGE YEO YONG-BOON

MACHIAVELLI[1] once said that men are held together not so much by love, which is fickle, but by fear. When the Soviet threat loomed large, the North Atlantic Alliance seemed eternal. China's "faults" were overlooked by the United States when it was needed as a strategic counterweight.

And with the old fear gone, feelings change dramatically. The North Atlantic Alliance has become less united. The European Community is being pulled in different directions. As for Sino-American relations, the love affair started by Richard M. Nixon (1913-1994) and Henry Kissinger has turned cold and is now replaced by a new fear that China may become a competing superpower in the twenty-first century.

Asean was formed in 1967 with Indonesia, Malaysia, the Philippines, Singapore and Thailand as its founding members.[2] It had

---

1 Niccolò Machiavelli (1469-1527), the Florentine philosopher, statesman and author of *Il Principe*, a treatise on statecraft, advocating the use of craftiness, deceit and unscrupulous cunning or guile to maintain political power and strengthen the State.

2 Asean is the acronym for the Association of Southeast Asian Nations. Brunei became a member of Asean in 1984 while Vietnam became a member in 1995.

a lackadaisical start until the pullout of American forces from South Vietnam in 1973 and its fall to the Communists in 1975. When Vietnam invaded Cambodia in 1978, Asean was galvanized into action. Without the Vietnamese threat, it is doubtful that Asean would have become the strong regional grouping it is today. It was cold fear, not love, which made them huddle together. Vietnam at that time had an armed force more powerful than the combined armed forces of Asean. To fight Vietnamese aggression, we created institutions to concentrate our efforts on the Thai-Cambodian border, in the United Nations and elsewhere. Without the challenge of Cambodia, there would not be this habit of consultation and goodwill we see today in Asean.

Whether Asean becomes stronger or weaker in the next century depends on the future strategic environment. Vietnam is no longer a threat. Indeed, by an irony of history, Vietnam now wants to be a part of Asean. What new fears will bind us together? In recent times, there has been a spate of optimistic write-ups in the international media forecasting a rosy picture for the region. If these pundits are right, the future of Asean would not be rosy for there would be little need for Asean. After all, Asean does not exist for itself but to serve the interest of its member countries.

Fortunately or unfortunately, there are new fears on the horizon, and some quite chilling ones. I am not being pessimistic about the future---indeed the future is full of opportunities and challenges for us---but there are also dangers we ignore at our own peril. The dangers are both economic and political.

New Fears

Whether or not we have a successful Uruguay Round on December 15, 1993, world trade will never be completely free. World trade will remain a mixture of free trade and managed trade. Of course, failure of the Uruguay Round will be disastrous, to say the least. It will slowly but surely divide the world into blocs. In that unhappy event, from our perspective in Southeast Asia, a two-bloc world, with the Asia-Pacific (including North America) as one large economic bloc and Europe as the other, is much better than a three-bloc world, with the Pacific divided down the middle.

If the Asia-Pacific region stays as one bloc, Europe will eventually be marginalized. The size of the Asia-Pacific region and the dynamism of many of its national economies will enable the region to grow regardless of European protectionism although everyone will be slowed down as a result. But Europe will lose much more and after some years will be forced to reopen its economy.

If the Pacific itself is divided as a result of an Uruguay Round failure, the consequences can be horrendous. We may still be able to muddle through. Or we may not. As international relations turn sour, trade wars can break out. A new great depression in the world is possible. Like the last one, a new great depression will trigger off a chain of political upheavals which may lead to another war. Such a scenario has been forcefully argued by Lord Rees-Mogg and James Dale Davidson in their sombre book, *The Great Reckoning*.

The success of the Uruguay Round is therefore a touchstone of mankind's ability to compete and co-operate within an agreed global framework. It does not mean, however, that we will then live happily ever after. Indeed, there will be fundamental shifts in the global balance of power which will lead to political conflicts of one kind or another in the future.

Among analysts, there is now broad consensus that, barring a major catastrophe, the world's economic centre of gravity will shift decisively to East Asia some time in the first half of the next century. There is every possibility that the size of China's economy alone will supersede that of the United States within two to three decades.

Such an economic shift has far-reaching strategic consequences. It will change the political and military balance of the world in a profound way. A transformation of such proportions has never taken place in history without accompanying wars and revolutions. Ever since the Spanish and the Portuguese expelled the Moors from Spain, successive European groups have exploded onto the world scene, each yielding power to the next only when forced to by defeat in war. Thus the Spanish and the Portuguese yielded to the Dutch, the Dutch to the British and the French, then the French to the British and, ultimately, after two world wars, the British to the Americans.

Japan, considered by the West at that time as an eastern upstart, was finally vanquished with the help of atomic bombs and only allowed to rebuild its economy as a defanged ally of the United States against the threat of world communism.

The last few centuries of Western domination has affected every aspect of our lives. The fact that we wear Western clothes, use Western toilets and kitchens, speak the English language, and apply Western ideas to all fields of human endeavour, is the direct result of the spread of Western civilization around the world.

Now, after five hundred years or so, the historical pendulum is inexorably swinging back to Asia, first to East Asia and eventually to South Asia as well. Just as the rise of the West was accompanied by innumerable conflicts amongst Westerners and between Westerners and others, the resurgence of the East will definitely be as tumultuous. Of course, one hopes that mankind has learnt from the lessons of history and will find ways to avoid major conflicts. After all, a major war in which nuclear weapons are exchanged will end civilization as we know it. But to assume that the swing of the pendulum back can be smooth is unrealistic and foolhardy. Such a hope flies in the face of human experience. It is somehow not in the nature of the species.

We are beginning to feel the rumblings. The gloom in Europe and the growing self-doubt in the United States are part of it. For Europeans to be talking seriously about reducing the work-week in order to share out the work available shows how deeply pessimistic they have become. The public mood is turning towards economic protectionism in order to save jobs. In the United States, H. Ross Perot's opposition to NAFTA tapped widespread grassroot support. Without Republican support, President Bill

Clinton would have had no chance of getting the Congress to pass NAFTA.

Anti-Asian sentiments are on the rise in Europe and North America. Domestically, there is growing racism against Asian immigrants because they work much harder and save more. Externally, an intricate theology is being constructed to justify why it is no longer possible to compete with Asians under the old rules. They claim that Asians have lower social standards. Asian workers are denied human rights. Prison labour is used. Their environments are being despoiled. The cohesiveness of many Asian societies is proof that they are authoritarian and do not practise genuine democracy. And so on. Thus an unwitting alliance of protectionists, socialists, human rights activists, environmentalists and liberal democrats is gathering in the West, all opposing free trade.

The arguments range from quite simplistic ones to the most sophisticated. H. Ross Perot's ability to simplify is well known. Samuel P. Huntington's seminal essay, "The Clash of Civilizations", clothes the race-protectionism argument in intellectual terms. Gore Vidal, the writer, puts it across very bluntly in an article in *The Sunday Telegraph* titled "Race Against Time". Vidal lamented the loss of "wealth, power and empire [which] makes for melancholy, or worse". The white men who made up 30 per cent of the world's population at the beginning of the twentieth century will be a "lonely" 13 per cent at the end of it. Vidal called for "a great northern peaceful economic alliance dedicated . . . to life, liberty and the pursuit of happiness". Not long ago, Jim Baker's "from Vancouver to Vladivostok" conveyed very much the same sentiment.

These are dangerous ideas which leaders like President Clinton and Chancellor Helmut Kohl are fighting against. We must encourage President Clinton in his support of APEC and a constructive Sino-American relationship. We in the East must be careful not to mismanage the growing insecurity of the West by adding fuel to the fire. In particular, we must never become anti-White ourselves. The EAEC idea was problematic because while many of us might have seen initial American opposition to it as being anti-Asian, many North Americans and Australasians probably saw it as being anti-White. The problem is not intellectual but emotional.

If the East and West retreat into racial positions, serious conflicts will be inevitable. It is far better for us if Americans and Europeans remain engaged in Asia and have a share in Asian growth and prosperity. Of course, we must expect them to exploit the natural divisions in Asia, and use Asians against Asians. Scholars like Chalmers Johnson recommend that America use Vietnam against China; Asean, China and Korea against Japan; Japan against China; and so on. They advocate balance of power realpolitik for economic advantage.

Indeed, from an Asian perspective, such a rational response by the West to the rise of the East is much better than an emotional response. There is then scope for bargaining and negotiation. But insecurity leads man to do foolish things. We cannot assume that the rational course will always be followed. Take China's recent bid to host the Olympic Games in the year 2000 as an example. Those who support human rights in China, the continuing liberalization of the Chinese economy and a through-train for Hong Kong in 1997 should rightly have supported Beijing, thereby holding Chinese policies hostage to the Olympic Games for years to come. But the mood in the West was anti-China. I told my Chinese friends that China was itself too eager and nationalistic. The photograph showing red flags fluttering on the Great Wall of China stretching as far as the eye can see was a bad mistake. Far from expressing China's welcome of the Olympic Games, which was the intended declaration, the display engendered deep fear instead. The red flags were an invitation to the proverbial bull.

Safety in Numbers

We cannot predict what the future holds but we can be sure that the journey there will not be smooth. Whatever the scenario, it is in our collective interest to huddle together in Asean. As Asean, we have a voice in the world. We have far more negotiating power as a group than we have as individual countries. If we are able to maintain the momentum of our economic development for another twenty years, Asean will become a significant player in the world economy in the next century.

We can only achieve such economic development if there is peace in the region and Southeast Asia does not get Balkanized. For every country in Asean, the pursuit of short-term national interest should be balanced against long-term collective interest. The Asean habit of *musyawarah dan musfakat* or "consultation and consensus" is thus very important. Indeed, the spirit of co-operation in Southeast Asia contrasts sharply with the antagonisms in Northeast Asia.

However, we must be realistic in our pursuit of consensus. We can never be as closely knit as the European community. For example, it is unlikely that member states will ever allow the free flow of workers across national borders. Nevertheless, we do share many things in common. From the cultural and historical viewpoint, Southeast Asia lies at the crossroads between two great civilizations: the Indian subcontinent and the Chinese mainland. For over two thousand years, the rise and fall of empires in South Asia and East Asia have left their marks on the society in Southeast Asia. Every country in Southeast Asia has been heavily influenced by Indian and Chinese cultures, not just the countries of Asean, but also the countries of Indo-China (hence the name) and Myanmar. The *gamelan*, for example, uses instruments which were originally Chinese to play music which were originally Indian, but it is now quintessentially Javanese and Balinese. Singapore, with over 77 per cent ethnic Chinese forming its population, has a name which needs no explanation to anyone from India. There can be no doubt that the economic resurgence of East Asia and South Asia will once again have strong repercussions on all of us in Southeast Asia, for good or for ill. If we were to recall the Chinese dynasties of yore, Han, Tang, Song and

Ming dynasties brought prosperity to the region but Yuan dynasty (under the Monggols) and the People's Republic of China (at the height of its revolutionary fervour) threatened everyone. However, the influence is not all one-way. Southeast Asia has also influenced developments in East Asia and South Asia. Rice cultivation, for example, originated in Southeast Asia. Today, Southeast Asia plays a significant role in the opening up of China.

The other fact of Southeast Asian history is that while the different cultures share much in common, they remain separate and distinct. Although Khmer, Thai and Burmese are Theravada Buddhists, they have fought bitter wars against each other for centuries. Malays and Javanese are different even though the majority of both profess Islam. Thus, our enthusiasm for Asean solidarity must be tempered by an understanding of the forces which divide us.

But because of the new fears which I described earlier Asean is more likely to strengthen than to weaken in the years ahead. A strong Asean enables us to treat with big powers like the United States, China, Japan, the European Community and India on a fair and equal basis. Divided, even the larger ones among us will be at a severe disadvantage.

Indeed, given the natural rivalry of the big powers, a united Asean enables us to extract much better terms for ourselves. In January 1993, Japanese Prime Minister Kiichi Miyazawa made an important policy speech on Japan-Asean relations. Praising Asean to the skies, he said that the *organis cohesion* of Southeast Asia was important to Japan and would be supported by the Japanese government. It would be surprising if Japan's support of Asean did not take into account Japan's relationship with the United States and China. In the United States, there is growing support for stronger economic ties between North America and Asean, the better to compete against Japan and China. Under President Bush, an Asean-United States Initiative was signed. Under President Clinton, the idea of a link-up between NAFTA and AFTA is now being floated. China and India, too, are wooing Asean.

Not to be left out, the European Community is also warming up to us. It is nice to be courted by all the major players. This happy situation is only possible because we stick together in Asean. Many foreigners do not understand the elliptical ways by which we reach consensus but that is our trade secret.

Part of our strategy must be to encourage the advanced countries to invest here and make Southeast Asia their manufacturing base. China is becoming a very strong competitor to Asean for foreign investments. In 1992, foreign investments in Indonesia, Malaysia and Thailand have all gone down as a result of better expected returns in China. Vietnam is also becoming a strong competitor. We will have to expedite the timetable for AFTA and expand its coverage if we do not want to be bypassed. The success of the latest Asean Economic Ministers' Meeting showed that, confronted with a serious challenge, Asean has the collective will to respond.

The goodwill that has been built up over the years in Asean is a precious asset. Here we must acknowledge the pivotal role played by Indonesia under President Surharto's leadership. Without Indonesia's early commitment to Asean, the history of Southeast Asia would indeed have taken a very different turn. With regional peace, our national economies, originally tied to the colonial powers, have become more intertwined. Japanese, American, European and, more recently, NIE investors, by their distribution of production facilities all across Asean, have integrated our economies in a complex way. Thus, what started out in Asean as an act of political co-operation against external dangers is now reinforced by cultural goodwill and economic integration. Official links are now buttressed by a multitude of cultural and commercial links. A clear sign of this is the dramatic growth in intra-Asean travel, both for business and pleasure.

Indeed, the success of Asean has persuaded the countries of Indo-China and Myanmar to seek eventual membership. It is only a matter of time before they become full-fledged members. For Vietnam, Asean is a way by which it can rebalance its historically difficult relationship with China. As for Myanmar, Asean offers a way out of its isolation and enables it to deal with both China and India in a more comfortable way.

Confronted with the same global uncertainties, Australia and New Zealand are also likely to move closer to Asean. Southeast Asia is after all Australasia's bridge to the Asian mainland. Large though it may be in geographical size and diversity, Australia has too small a population for it to be really effective on its own in world politics and in trade negotiations. Australia's first preference is of course for an Asian-Pacific community in which no hard choices are necessary. In a storm, however, Australia may have to choose with whom it hunkers down, whether with North America with which it shares a common Anglo-Saxon and Celtic heritage or with Southeast Asia, its immediate neighbour. Intellectually, the choice would probably be Asean. Emotionally, it is much less clear. However, a recent survey showed that the majority of Australians still do not consider themselves part of Asia. Nevertheless, this orientation will gradually and surely change as the Australian and New Zealand societies become more Asianized. From the perspective of Asean, it is very much in our long-term interest to forge stronger links with Australia and New Zealand. It is also important to demonstrate that we in Asean are not racist even though others may be.

Conclusion

We do not know what the world will be like ten or twenty years from now, only that it will be very different from what it is today. Straight-line projections are of no use. There are incredible opportunities before us but also great dangers. Because of the shift of power across the Pacific, conflicts of one kind or another are bound to occur. We will be dragged into them regardless of what we do. It is therefore very important for us in Southeast Asia to raft our boats together. We are then more likely to stay afloat when storms break. If the Uruguay Round succeeds on December 15, 1993, we can breathe a deep sigh of relief, at least for a while. But the long-term uncertainties remain. No nation gives up power or property without first putting up a fight. If the Round fails, events may move quickly.

Asean's strategy in this period of uncertainty should be twofold. First, we should broaden our base by gradually bringing in Indo-China and Myanmar as members, and Australia and New Zealand as close partners. Second, we should deepen Asean integration through AFTA. Only then can we deal with the big powers as an equal and retain some control over our destiny.

3

Opportunities for the Transfer of Technology in the Pacific Rim

WILLIAM F. MILLER

THE growing importance of technology in the world economy and how that has affected technology flows and technology transfers may be examined by looking at its historical perspective.

There are two forces at work that strongly influence technology development and technology flows:

1. Increasing contribution to the growth of the world economy from knowledge-intensive industries and services; and
2. Increased global competition and the drive for productivity worldwide.

A brief examination of global trends in per capita growth of GDP shows that the contribution to this growth from basic raw materials has tapered off or declined, in some cases while the aggregate world GDP per capita continues to increase. At least for the foreseeable future, the increases in aggregate worldwide growth per capita of GDP will come from the knowledge-intensive industries and services—computers, electronics, pharmaceuticals, chemicals and new materials. Countries or regions within countries that do not play some role in these knowledge intensive industries, will not participate in the growth of the world economy. This lesson has not been lost on the developing countries of the world and especially has been clearly recognized by the Asian countries.

This does not mean that agriculture and natural resource industries are unimportant. Indeed, agriculture and natural resource industries can make important contributions to a country's economy and a country's wealth generation. With appropriate macroeconomic policies, agriculture and natural resources can be used to fuel the growth of the industrial economy as was done in Taiwan and the Peoples' Republic of China in the 1970s and 1980s.

Increasing knowledge intensity means increasing use of technology. Technology is the embodiment of knowledge. More and more knowledge is being embedded in technology, both hardware and software. The highest value added products are those with the most knowledge embedded in them---for example, smart cards, smart telephones, smart glass and smart weapons.

This development has created a worldwide search for new technologies that can be effectively used to improved products, as well as production processes. Nowhere is this search more intense than in Asia. The Newly Industrialized Economies (NIEs), which for two decades or more acquired lower level technologies and capitalized on low labour costs, have begun their move upscale as have other Asian nations. The earlier strategy was so successful that the NIEs are no longer low labour economies so they are offshoring their low labour cost work, licensing higher level technologies when they can, engaging in various alliances that give access to technology, and increasingly producing their own.

As a means of producing their own technologies, Asian countries are rapidly increasing their expenditures on research and development (R&D), both by the public and private sectors. The patterns of expenditure, however, differ from country to country. In terms of expenditures on research and development as a percentage of GDP, the United States and Japan are still considerably ahead of the NIEs and China, however these countries are moving to close the gap. In Japan and South Korea, industry is by far the largest contributor of funds for research and development. In India and China, the government is the biggest contributor, whereas in Singapore, Taiwan and the United States, there is nearly an equal balance, with industry contributing slightly more than government. In India and China, the government is still the largest performer of research and development with growing participation by industry, whereas in all other countries, industry is by far the largest performer.

Of the Pacific Rim countries surveyed,[3] the United States' expenditures for research and development have been growing the slowest, about 4 per cent per year in the last fifteen years, however, as a percentage of GDP, they have grown from 2.2 per cent in 1975 to 2.7 per cent in 1990.[4] Over that same period of time, Japan increased its research and development expenditures as a percentage of GDP from 1.8 per cent in 1975 to 2.9 per cent in 1990. In the last three years, there has been a slowing down and even cut back on research and development in Japan as companies were feeling the effects of the recession and industrial restructuring. In other Asian countries, the growth of research and development spending ranged from 10 per cent per year in India to 23 per cent per year in South Korea.

The conclusion one draws is that we see very substantial emphasis on research and development in the science and engineer-

---

3 "Human Resources for Science and Technology: The Asian Region," Special Report, National Science Foundation NSF 93-303, Washington, D.C. 20550, 1993.

4 R&D fared well in the United States' 1994 federal budget. The National Science Foundation received a 10.7 per cent increase over 1993, a 35.4 per cent increase for the National Institute of Standard and Technology, and a 6.1 per cent increase for the National Institute of Health.

ing areas in recognition of the importance of "home-grown" technologies.

The other side of this coin is that research and development has become more expensive. research and development spending as a percentage of sales in the knowledge intensive industries varies from 5 per cent in communications to 15 per cent in computers. Within an industry research and development spending as percentage of sales may vary considerably. For example, in the computer industry, the semi-conductor segment spends close to 20 per cent of sales on research and development.

At the same time that we see growing importance of research and development, we also see cost pressures on companies to respond to the intense global competition. Worldwide we see companies downsizing and restructuring to become more cost competitive. Industrial research cannot avoid these same pressures. While company-funded industrial research and development in the United States increased from 2 per cent of sales in 1975 to 3 per cent in 1990, in the last few years the rate has not increased and in some industries it has decreased.[5]

With increasing importance of research and development, increasing costs, and at the same time global competition increasing the cost pressures on companies, efficiency in use of research and development and of technology gets increased attention.

Asian countries are now better prepared to make efficient use of technology including technology transferred from the United States or Europe, as well as technology from other Asian countries and that developed at home. The principal enabler of this is the increased numbers of science and engineering employees in the work force. All Asian countries are increasing the science and engineering employees as a percentage of the work force by sending large numbers of students abroad for advanced degrees, in-

---

[5] "The Competitive Strength of United States' Industrial Science and Technology: Strategic Issues," *National Science Board Report*, August 1992.

creasing the numbers of science and engineering graduates at home, and enticing expatriates abroad to return home for work in both universities and industry. In order to efficiently transfer technology, it is necessary to have well trained scientists and engineers who can receive the technology. All Asian countries surveyed are positioning themselves to become better recipients and users of technology by indigenous companies.

Foreign direct investment (FDI) in developing Asia is increasing rapidly. There is a very rapid increase of intra-Asian investment, but still the major FDI in developing Asia will come from the United States and Japan. For example, the United States increased its FDI flow to Asia by 120 per cent between 1989 and 1992. These increased flows of FDI generally mean increased flow of technology as well.

The prospects, then, for increased technology transfer in Asia seem generally bright—stronger indigenous research and development; stronger cadres of scientist and engineers; more FDI, especially in science and technology; and a growing prosperity and markets on the demand side. But there is one remaining issue that has to be settled—intellectual property protection. The problem is especially serious in the software area. More and more knowledge is being embedded in software. Often the software associated with products is the highest value added part of the product. Yet the protection of software remains a sticky problem. All Asian countries have moved to strengthen their laws and regulations for the protection of intellectual properties including software. The problem apparently is not the existence of laws and regulations, but one of enforcement.

This is not a United States versus Asia issue. This is an Asian versus Asian issue. I am the director of a Computer Industry Project at Stanford University. As a part of that project we are studying the software industry in Europe, North America and Asia. Let me illustrate some of the problems of software protection. The head of a Japanese software reported to one of our researchers that his biggest problem was unauthorized copying of his software. The laws in Japan are adequate for protection of software. However, both private companies and government agencies routinely make multiple unauthorized copies of software.

In another country, the manager of a software department reported to one of our researchers that they would like to make investments in another country, but they are deterred from making those investments because of rampant software piracy, that is, they fear that their software may be copied, distributed and sold.

The problem seems to be that there is a very low regard for software. Many people see software as a public good or something of low value. There is not the recognition of high intellectual content and the embodiment of knowledge in the software. As Asian countries move up the value-added scale in technology, software will become a more important part of their products and indigenous companies will need better software protection and especially better enforcement of existing laws and regulations. This will become increasingly more important to Asian companies if efficient technology transfer is to be achieved.

4

Multicultural Management: New Skills for Global Success

FARID ELASHMAWI

WHAT should you do if you disagree with your American manager? How would you criticize Taiwanese workers? What subject should you bring up first in business discussions with the Chinese? How would you begin your presentation to an Arab company?

These and other questions are typical of the intercultural encounters managers must face in this era of globalization. Depending on their expertise, in such situations, managers can affect the outcome of overseas joint ventures.

Companies often venture overseas nowadays. Sometimes, governments urge them to do so. At other times, economic imperatives dictate expansion abroad; escalating costs of doing business at home make it worthwhile to explore overseas markets. Companies can only remain competitive by going global as dependence on the domestic market is no longer lucrative or feasible due to limited possibilities for growth and expansion at home. To compete in the global marketplace, many organizations may have already overcome the political, technical and financial barriers to opening up markets overseas. But most organizations are not yet able to fully cope with the issues of multicultural management.

This chapter examines the sources of friction caused by cultural differences in many joint ventures and is based on the author's development of, and experience with, multicultural training programmes for managers. The programme explores how to apply new skills in interpersonal business introductions, multicultural meetings, business presentations across cultures, training and motivation of foreign workers, and negotiations.

What is culture? Is it art? Language? Religion? Pattern of thinking? Values? How would you answer this question? Most managers in the United States and other western countries say that culture is beliefs, values, ways of thinking and language. The Japanese and most Asians mention history, tradition and ways of life. Arabs agree with the Japanese, and Asians add religion. These culturally conditioned responses are clues to variations from culture to culture.

We can define culture as a behavioural norm that a group of people have agreed to survive on. These norms vary by time and place and are constantly adapted to the changing environment. Within each culture there are many subcultures of which an individual is simultaneously a member. An American manager, for example, might belong to these subcultures: Polish-Americans, IBM, sales department, product group, family. Each of us represents an individual subset of cultural values.

What do you value most? Competition? Formality? Group harmony? Risk-taking? Authority? In my experience with a diversity of cultures, the Americans consider independence, competition and individual success the most important values, the Japanese, group harmony, co-operation and group achievement, while the Arabs consider family security, compromise and reputation. Once we have explored the wide definition of culture and identified our own values and how they differ from other cultures, we begin to realize that all our behaviour in life and business is culturally oriented. This realization becomes a tool for comprehending why that which is normal in other cultures may look abnormal to us.

Each of us behaves in a way that is normal and consistent because that is the way we grew up, and our behaviours are those we have been rewarded for. However, for people of one culture to communicate and manage people from other cultures, they must first understand how their own values may conflict with those of other cultures.

Normal, day-to-day business activities are affected by cultural differences. These activities include personal introductions, meetings, presentations, training, motivation and telephone or written communication across cultures. For example, if you disagree with your American manager, what would you do? Would you:

1. keep quiet because you shouldn't challenge the boss;
2. talk with your manager after work;
3. talk to your manager's boss about your disagreement; or
4. openly discuss the matter directly with your manager?

Most Asians will choose the first option because Asian culture accords great value to seniority. The Japanese value group harmony and will probably select the second option. Arabs will choose the third option because in Arab culture, an employee who cannot solve a disagreement with his immediate supervisor may decide to appeal to a higher authority while Americans will choose the fourth option because they value equality, and will most likely deal directly with their managers.

Without an understanding of other people's values, many workers and managers will define everyone's behaviour according to their own cultural values. This is the source of cultural conflicts. For example, in a business discussion, an Arab executive may stand closer to you than most Americans or Japanese would. This means that the Arab:

1. wants to tell you something confidential;
2. wants to smell your new cologne;
3. is expressing his hospitality; or
4. does not want anyone else to join in the conversation.

In such a situation, most Westerners and Japanese will retreat because they may feel their Arab counterpart has invaded their comfort zone of personal space. This retreat will send the wrong message to the Arab executive who is standing at a distance that expresses personal warmth and hospitality for the business associate. Without recognizing cultural differences, managers will expect actions based on their own cultural experience.

During business meetings, the Japanese listen with interest but do not make decisions. This is because their culture values group consensus, and they will follow the meeting with discussions of each point with each person present to determine how to respond.

Japanese and Arab cultures hold a much longer perspective on time than Western cultures. They consider the time spent building a relationship of mutual trust part of the cost of doing business.

When working with or training Arab workers, criticizing them in front of their peers will make them lose face and damage their relationship with you. Arabs are usually coached on a one-on-one basis by their immediate supervisors. And Chinese business discussions usually begin by mentioning mutual relationships and co-operation to lead to more open-door talks.

When making a formal presentation in South America, it is expected that you will impress your counterpart by stressing your company's reputation and your own influential position in the final decision-making.

What we have learned with our multicultural experiences in business is that to start communicating successfully on a global scale, you have to begin with yourself and your own value system. You must understand how your value system is different from that of others and why. Only then can you proceed with the notion that every culture is right and that the key to success is interculture.

Test your global communication and management skills by reading the questions below to find the most appropriate answer to each. Identify the cultural value behind each of your answers. How might your answer be different from that of a person of another culture?

1. You are making a presentation to a group of executives from the United States to get their support to go ahead with your project. They will be most impressed by your:

   a. graduate courses and names of the schools you attended.
   b. current level of responsibility.
   c. record of past accomplishments.
   d. management style.

2. You have just received a fax from an Arab dealer asking you for a quote for 10,000 computers. He is really interested in:

   a. 10,000 computers.
   b. your ability to deliver a large order.
   c. 10 computers.
   d. your bottom price.

3. Betty Hu, from Taiwan, is new on the assembly line, and she is still having some difficulty putting the units together properly. You want to train her. You should:

   a. tell her, "Betty Hu, you are learning fast," then show her exactly how the units are put together.
   b. say, "Betty, you should pay closer attention to what you are doing."
   c. shout across the room, "Betty, do you need help?"
   d. take her aside and show her again how to put the units together.

4. You are sending a new piece of equipment to Japan. Which of the following is most important to send with it?

   a. a detailed technical description in English.
   b. a detailed technical design manual in Japanese.
   c. audio and video tapes in Japanese on how to use the equipment.
   d. a technician to explain how to use the equipment.

5. You are entering a joint venture with a local Chinese enterprise. You want to hire a local general manager. You should select him on his past ability to:

   a. improve profit by cutting staff.
   b. get money from the government.
   c. manage his organization.
   d. learn new technology.

6. As European General Manager of a production factory in Singapore, you decided to have an open-door policy, but no one came in. They reason may be:

   a. workers will not go over the head of their first line manager.
   b. the workers do not like you.
   c. it is impolite to meet with the director without an appointment.
   d. the workers like to work as a team and do not care much about the boss.

7. Your company has been invited to discuss a high-technology project with officials in the USSR. Their primary concern in the first meeting will be:

a. profit margin.
b. delivery schedule.
c. technical specifications.
d. who will fund the project.

8. You have been invited to give a training seminar in Mexico. They ask for your fee and terms. Do you:
   a. quote a lower fee since you have never been there before.
   b. ask how many participants there will be and what they are paying per person.
   c. quote a lower fee but insist on first-class air travel.
   d. double your normal fee to leave room for negotiation.

Most appropriate answers: 1-c, 2-d, 3-a, 4-d, 5-b, 6-a, 7-c, 8-b.

5

Partnership for Profit: Managing Strategic Alliances

ODED SHENKAR

IT is my firm belief that the importance of strategic alliances goes far beyond the domain of business. In today's constantly changing world, in this era of opportunities (and also suspicion and threats), international strategic alliances represent many hopes: hopes for co-operation, hopes for understanding, and hopes for learning to trust and learn from each other. Possibly the first time in recorded history, international business can be conducted not through dominance, but through co-operation; not through confrontation, but through friendship. And this opportunity presents itself precisely when global business has become such a focal point for international relations. Therefore, our hopes are for strategic alliances becoming a vehicle for a better world.

The increase in the number of strategic alliances is staggering, to say the least. Joint ventures, a major subspecie of strategic alliances, have become a major route for internationalization. The number of foreign joint ventures registered in China in 1992 alone exceeds that of the previous twelve years combined. In the same year alone, more than twelve thousand foreign joint ventures were registered with the Hungarian authorities (a cumulative figure). Other countries, both in Asia and elsewhere, paint a similar picture. However, I would like to bring into focus a little-known case: Hong Kong. In this vibrant British colony, perhaps the purest free market we know on this planet, there are no legal or other limits on establishing wholly-owned foreign subsidiaries. Yet, when we looked at the foreign investment figures in the manufacturing industry in Hong Kong, we found that about half the companies surveyed opted to establish a joint venture with a local company. Apparently, many firms are no longer dragged into joint-venture arrangements against their wishes; rather, they make a conscious decision to co-operate with local firms.

This brings us to another key question: why do firms form international joint ventures (IJVs)? I do not intend to tire you with a detailed depiction of the economic theories on foreign direct investment (FDI), from market power to transaction costs (the latter, for instance, suggests that firms will opt to exchange goods and services in the market or produce them alone depends on the relative effectiveness of each mode, joint ventures being a "mid-point" on this continuum of options). I will only say that it is my strong belief supported by both scientific research and practice in the field, that these models are insufficient. They are based on long discredited assumptions on the rational nature of company decision-making, and they are tested by submitting existing venture distributions to a post-mortem analysis which excludes all aborted attempts to establish such ventures. None of these models, for example, could predict the flow of Taiwanese investment to China's Fujian's province, or explain why Americans of Hungarian ancestry have led the way in US investment in Hungary. None of these models would explain why AMC, the smallest of American car manufacturers and now part of Chrysler Corp, was the first to set up an automobile joint venture in China.

What has been missing in the conventional economic models is the human element, the individual's curiosity, interests, preferences, and, yes, his career prospects. Most of all, his ability to create "chemistry"----to establish the relationships that are critical not only to the establishment of such ventures, but also to their successful operation once formed.

Over the years, I have observed many joint ventures, and also many "divorces". In a great number of cases, divorces were not only attributed to strategic on cultural misfit—topics which I will explore later—but rather to a break in the relationship, for instance, when a key member, a "secondary-spanner", in one of the parties has been replaced, thus undermining the vital relationship linking the parties together.

Generally speaking, we have paid much more attention to the supposed strategies for IJV formation than to their actual management. And yet, the two are inseparable. If the joint venture is a marriage—a question which we will tackle later on—it is reasonable to assume that divorce has to do with the dynamics of married life than only with initial compatibility. If IJVs contain "self-destruct devices" as Young and Bradford suggested over two decades ago, then we must look for ways to diffuse those devices before they tear the venture apart.

Doing so requires an understanding of the elements of human resources: the unprecedented variety of employee groups and the national and corporate cultural cues they bring with them. For example, while wholly owned subsidiaries have three employee groups, foreign expatriates, host country nationals, and third country nationals, joint ventures have up to eight. We should also familiarize ourselves with the typical human problems in such ventures, from staffing friction through compensation gaps to split loyalty; and from blocked promotion to the screening of information with other problems related to the reality of IJVs as multipartite, international entities whose parents' objective may not always coincide.

Remember the value—but also the limits of legal advice and detailed contracts, by themselves the subject of cultural contention between the East and West. Lawyers in the United States who specialize in prenuptial agreements between couples will tell you that a significant portion of couples contemplating such agreements decide, midway through the process, to call off the marriage. Better that than a divorce, you might say, but you will also be missing, perhaps, on that suitable partner who has been turned off by your mistrust. And no contract will ever ensure the success of married life.

If I have frightened you from courting potential partners here, it would have been contrary to my intention. The challenge of forming and managing strategic alliances is tremendous—but so are the rewards to the companies, individuals, and, yes, the nations involved.

Joint ventures have been frequently compared to marriages. Though the term may be more suitable for mergers, where partners converge all their assets, it is also a term useful for the understanding of joint-venture establishment and operation. Indeed, while the number of joint ventures has skyrocketed, so has the number of divorces. And while some divorces have been amicable with the partners praising "the wonderful years together", (translated to joint-venture language as "we have each achieved our respective objectives"), others are not. In both the human and the corporate world, a divorce may be costly, both materially and mentally. Having "burned their fingers" some partners decide not to try again.

It is therefore no wonder that for both human and firm marriages, it is frequently noted that "partner selection is the most important". After all, you don't just decide to marry, you rather decide to marry a particular individual, or, in our case, a company. Yes, a "matchmaker" may be involved, but ultimately it is you who will have to live with the entity you have chosen.

It is here that we should note that joint venture marriages are typically more complex than their human equivalent. Many firms have multiple partners (Corning Glass, for instance, has almost thirty), who may resent the attention each other gets, yet are tied together in an elaborate network we call "a system". The terms of the marriage, then, are more critical in joint ventures. How long is the duration of the venture? What are the dependencies created among the various system components (for instance, is the venture dependent on one of the partners for essential inputs)? These are only samples of the questions you need to ask yourself when considering a joint effort. It is very important that you ask these questions not only in relation to your own firm, but also in relation to your potential partner. As in the human case, complementarity may be more important than similarity. Coming from Hawaii, the only state in the United States where the number of marriages across ethnic lines is actually higher than the number of marriages within the same group, I feel qualified to talk to you not only about the complexity of such marriages, but also about the happiness they can bring. It is your role to ensure that the same is true for your respective companies.

Know your options: determine which form of strategic alliance is right for you, and whether this is the form suitable for your partner. Ask yourself what you expect the alliance to do for you, what it should do for your partner, and whether the two are compatible. Should you decide on an equity joint venture, consider teaming with more than one partner—contrary to conventional wisdom, many CEOs tell us that multi-parent ventures are easier to manage than the two-parent variety. Judge what equity distribution is right for you, and how comfortable you are with a limited control over operations. Determine which partners to seek, and ask yourself to what extent you have a "fit"—strategically, culturally, and interpersonally. Most importantly, be open, and understanding—"This may be the beginning of a beautiful friendship"—but, if you don't work on the relationship, it may not last much longer than the shifting alliances, so aptly described in the Hollywood classic, *Casablanca*.

The most recent studies on human marriages, namely, the one by Gottman and Markman, concluded that there are four behaviours which, taken together, will likely undermine a marriage: criticism, contempt, defensiveness and withdrawal. Though these symptoms have not been studied systematically for joint ventures, my experience suggests that they are equally important in that context. Respect for differences, open communications, and the willingness to engage in an open and constructive debate which recognizes diversity in both the content and style of discourse, will make it more likely that your partnership will prosper, and that the company or "child" you have created (in the case of joint ventures), will bring forth not only profits but also pride and joy to the parents.

6

Malaysia: Asia's Profit Centre

MALAYSIAN INDUSTRIAL DEVELOPMENT AUTHORITY

MALAYSIA aspires to be a fully developed country by the year 2020, developed not only economically, but also politically, socially and spiritually. Priorities have been set and the development path mapped out by the government to make sure the objectives of Vision 2020 are met. The manufacturing sector is expected to spearhead the drive to fulfil this vision. The challenge will be for the private and public sectors to work together to establish a strong national entrepreneurial base that will carry the nation forward into the twenty-first century. The main thrusts would be to:

1. Develop a strong science and technology base to enhance industrial competence and competitiveness;
2. Develop indigenous technology by increasing the nation's capability to adopt, adapt and improve technology through research and development;
3. Develop a pool of skilled manpower capable of handling new and emerging technologies;
4. Ensure that environmental conservation and protection are built into the overall economic development process; and

5. Develop a positive culture based on integrity, discipline and diligence to equip Malaysians for high productivity and commitment to quality.

Malaysia: Your Manufacturing Base in Asia

Malaysia launched into manufacturing in the 1960s with a range of import-substitution industries. Today, the nation has progressed into the export-orientation era and is recognized as one of the world's leading exporters of electronic semiconductors, room air-conditioners, audiovisual equipment, and products based on her wealth of natural resources such as rubber, palm oil and timber. According to the GATT (General Agreement on Tariffs and Trade), in 1992 Malaysia joined the ranks of the world's top twenty-five merchandise exporters.

Malaysia ranks high among investors as an attractive location for their offshore operations. According to the World Competitiveness Report 1992 published by the World Economic Forum in Geneva, Switzerland, Malaysia ranks fourth among Asian countries in terms of international competitiveness (right after Singapore, Taiwan and Hong Kong). Among the eighteen thousand senior executives surveyed, Malaysia was accorded high marks in the areas of basic research, technological advancement, long-term competitiveness and hard work.

Malaysia's business environment is enhanced by her strategic location within the heart of Southeast Asia which positions her to serve the growing markets of Asean and the newly emerging markets of China, the Indo-China States and Myanmar. The Asean economies, with a total population of almost 330 million, enjoy growth rates ranging from 5 per cent to 8 per cent. Based on present growth rates of these economies and their populations, the GNP for the Asean region is projected to reach US$500 to US$550 billion by the year 2000. The tremendous potential generated by this growth makes it imperative for manufacturers to locate their operations within the region.

Asean will become an even more attractive market with the establishment of the Asean Free Trade Area (AFTA). The Agreement on the Common Effective Preferential Tariff (CEPT) for the AFTA which liberalize trade within Asean through the gradual reduction of tariffs, aims to establish a free trade area within fifteen years, commencing January 1, 1993. By establishing a production base in Malaysia, manufacturers will be able to export freely or within minimal import duties to other Asean member countries.

A Policy of Welcome

Malaysia welcomes foreign investors in her manufacturing sector and adopts a global approach in her drive for industrial development. The government sees the liberalization of the economy as the natural path for the country to take so as to compete in the international marketplace.

An outstanding factor in Malaysia is the government's sincere commitment to creating an investment environment that will provide businesses with the opportunities for growth and profits---a factor which has persuaded more than three thousand international companies from over fifty countries to make Malaysia their offshore base.

Malaysia's welcome to foreign investors is also seen in the establishment, since 1972, of overseas offices by the Malaysian Industrial Development Authority (MIDA) in capital and technology exporting countries. MIDA is Malaysia's principal agency for the promotion and co-ordination of industrial development in the country.

Malaysia has hitherto signed Investment Guarantee Agreements (IGAs) with more than twenty countries. The IGAs ensure that there shall be no expropriation or nationalization without adequate compensation, and allow for remittance and repatriation of capital and profits. According to the World Competitiveness Report 1992 published by the Geneva-based World Economic Forum, Malaysia gets top marks among the newly industrialized economies in terms of investment protection schemes available for foreign partner countries.

Kuala Lumpur is also the base for the Regional Centre for Arbitration, a non-profit administering agency whose main function is to serve the needs of international organizations and business communities in the Asian and Australasian region for reliable international arbitration in settling commercial disputes.

Equity Participation

Foreign investment in the manufacturing sector has been greatly encouraged by Malaysia's liberal policy on equity participation which allows up to 100 per cent foreign ownership in projects which export 80 per cent or more of their production. For projects which export less than 80 per cent of their production (down to a minimum of 50 per cent export), up to 100 per cent foreign equity ownership is allowed when:

1. Foreign investment in fixed assets (excluding land) is at least RM50 million or when the project has at least 50 per cent value added; and
2. The company's products do not compete with products presently being manufactured for the domestic market.

Expatriate Posts

Expatriates are welcome to work in the country only when there is a lack of trained Malaysians to fill the positions. In addition, key posts are allowed in order to safeguard the interests of foreign investors. Key posts are positions which can be held permanently by expatriates.

Exchange Control

Exchange control regulations have always been liberal in Malaysia. All payments abroad, including repatriation of capital and remittance of profits, are freely permitted. In addition, Malaysia's accession to Article VIII of the International Monetary Fund's

Articles of Association assures free transfer of funds and free interchangeability of currency.

Tax Incentives

The private sector complements the government in spearheading the development of the Malaysian economy. This Malaysia Incorporated approach has created a supportive environment that encourages efficient communication between the government and the private sector.

While the private sector is expected to hold the lead role in achieving the nation's overall development objectives, the government on its part will create a conducive investment climate to bring out the resourcefulness and dynamism of private enterprises such as the provision of a range of tax incentives structured to encourage self-reliance and the proliferation of competitive and efficient industries.

Malaysia's principal tax incentives for the manufacturing, agricultural and tourism sectors are contained in the Promotion of Investments Acts, 1986 and the Income Tax Act, 1967. They include the Pioneer Status and Investment Tax Allowance which grant relief from taxes and are provided to companies undertaking new projects, expansion or diversification. Other incentives in the form of tax deductions are also provided to encourage exports, research and development and training. Companies setting up operational headquarters (OHQs) in Malaysia will also be taxed at concessionary rates.

Infrastructure

Malaysia's developed infrastructure is one of the main attributes that have placed it among Southeast Asia's most rapidly industrializing nations. New initiatives are constantly being planned and implemented to upgrade infrastructure development in line with the country's economic growth, while privatisation has provided for increased efficiency in the utilities sector.

Malaysia believes in investing in infrastructural facilities to provide a conducive environment for investors. A substantial allocation of RM26 billion has been set aside to finance the upgrading of transport and communication infrastructure during the Sixth Malaysia Plan period from 1991 to 1995.

1. **Good Roads**
   Malaysia's highway network is the backbone of the country's transport system: 90 per cent of passenger and cargo movements is by road. More than 40,000 kilometres of well-maintained roads link all parts of the country. This is complemented by railways which traverse the west and east coasts of Peninsular Malaysia.
2. **International Airports**
   Malaysia's five international airports: Subang-Kuala Lumpur, Penang, Senai, Kota Kinabalu and Kuching, are serviced by most of the world's major airlines. Today, the rapid expansion of trade and industry, together with a growing tourism industry, have created a need for more international airports. Currently, a new and highly sophisticated airport is being planned in Sepang, Selangor with a capacity for sixty million passengers per annum. This airport is targeted for completion by the end of 1997.
3. **Well-Equipped Seaports**
   Of the five major ports in Malaysia: Port Kelang, Penang, Pasir Gudang and Kuantan in Peninsular Malaysia and Bintulu Port in Sarawak, the ports in the peninsular are well-equipped to handle international container traffic, while the Bintulu Port caters mainly to the LNG industry.

Modern Telecommunications

Malaysia's telecommunications network has seen impressive expansion and upgrading especially over the last five years, and is now one of the most well developed among the industrializing countries of the Asia-Pacific. International communications have been facilitated by the introduction of state-of-the-art technology in telephone, telex, telefax and videoconferencing services, as well as other data transmission and networking facilities. The latest digital and fibre-optics technology is now being deployed in efforts to provide high-quality telecommunication services at affordable prices.

Well-Developed Industrial Parks

Industrial estates, free trade zones (FTZs) and licensed manufacturing warehouses (LMWs) are provided in Malaysia specifically to attract investment. Suitable sites for industries are provided by more than 170 industrial estates or parks located throughout the country. These industrial sites, designed for manufacturers producing or assembling products mainly for export, are equipped with basic infrastructural facilities such as roads, electricity and water supplies, and telecommunications. To cater to the needs of export-oriented industries, export-processing zones (EPZs) called free trade zones have been developed where customs controls and formalities are reduced to the minimum. Companies in FTZs are allowed duty-free imports of raw materials, components and parts, machinery and equipment directly required in the manufacturing process. Similar facilities are also provided by licensed manufacturing warehouses. These are similar to FTZs but smaller and designed to encourage the dispersal of industry througout the country.

Specialized industrial parks have been and are being developed to cater to the needs of specific industries. Examples of these parks include the Ceramic Park and the Pengkalan Foundry and Engineering Park in the state of Perak, the Olak Lempit Furniture Complex and the Selangor Marine Industrial Park in the state of Selangor, and the Kulim High-Tech Park in the state of Kedah.

Human Resources

One of Malaysia's key attributes is her people. They make up a pool of young, educated, trainable and highly productive workers. The country's labour force comprises 7.5 million people, 60 per cent of whom are below thirty-five years of age. Wage rates are competitive compared with those in other countries in the region.

About 950,000 students enter secondary schools every year in Malaysia. They complete lower and upper secondary education with a total of nine and eleven years in school respectively.

Eight universities and numerous technical and vocational institutions provide a stable supply of trainable labour to support industrial growth. An estimated 173,000 students at the degree, diploma and certificate levels will graduate from local institutions during the Sixth Malaysia Plan period (an average of 49 per cent from science and technical courses), while another 50,000 Malaysian students are enrolled in tertiary institutions abroad. During the Sixth Malaysia Plan period, public industrial training institutions will increase their intake to produce more and better trained manpower to meet the requirements of industries. About 197,000 trainees, an estimated 70 per cent in engineering trades, will complete training programmes at the skilled and semi-skilled levels.

The government has also allocated a sum of RM48.9 million to be used over a three-year period beginning in 1993 for the Human Resources Development Fund. Manufacturers who are contributing to this fund will be eligible to apply for grants under the fund to defray or subsidize costs incurred in the training of their workers, especially in new and emerging skills.

To ensure industrial harmony, Malaysia's labour laws set out minimum standards to safeguard the interests and spell out the rights and responsibilities of employers and employees. These provide a legal framework for the orderly conduct of industrial relations in the country.

The Business Environment

International corporations are comfortable doing business in Malaysia's comprehensive and well-developed business infrastructure where English is widely used. Legal and accounting practices in Malaysia are derived from the British system and would be acceptable to most companies doing international business.

The government's Malaysia Incorporated approach provides effective channels of consultation between the public and private sectors. This can be seen in the many business organizations in Malaysia which have regular dialogues with the government—the Malaysian Business Council, the Malaysian International Chamber of Commerce and Industry, the Federation of Malaysian Manufacturers and the National Chamber of Commerce and Industry, besides various foreign business associations. Newcomers to Malaysia will be able to get assistance from these business associations.

A well-developed financial and banking sector has also enhanced Malaysia's position as a prime investment location and a dynamic export base in Asia. Sophisticated financial facilities are available through thirty-seven domestic and foreign commercial banks (with a nationwide network of over 1,125 branches), seven industrial finance institutions, twelve merchant banks, forty-one finance companies and seven discount houses as well as other financial institutions. There are also about thirty representative offices of foreign banks. Commercial banks, finance companies, merchant banks and industrial finance institutions are the major sources of credit to the industrial sector in Malaysia.

In 1990, to complement the existing financial system in Kuala Lumpur, an international offshore financial centre (IOFC) was established in the federal territory of Labuan, off the north-west coast of Borneo. Labuan aims to be a major centre for offshore banking, insurance and trust activities in the Asia-Pacific region.

The Kuala Lumpur Stock Exchange (KLSE) has evolved into one of the world's foremost emerging markets, supported by rapid economic growth, strong company profitability and marked increases in foreign capital inflows. The United States-based International Finance Corporation (IFC) has rated the KLSE as one of the top five stock exchanges among the world's emerging markets in terms of market capitalization. The Kuala Lumpur Commodity Exchange (KLCE) continues to introduce new measures to increase the volume of trade and liquidity of its existing futures contracts.

In industries, Malaysia offers manufacturers the competitive edge in various sectors. These include:

1. Resource-based industries, that is, industries which utilize Malaysia's abundant resources such as rubber, palm oil, tin, oil and gas, cocoa, silica sand and clay, to manufacture high value-added products, especially those for export markets.
2. Industrial machinery and parts, machine tools and related engineering industries.
3. Ancillary and supporting industries, such as the mould and die industry and industries supplying parts and components, which provide the important linkages that support the established larger industries.
4. Precision products such as medical instruments, musical instruments, photographic and optical goods, watches and clocks.
5. Agro-based industries such as the cultivation and processing of fruits, vegetables, livestock and aquatic products.

Manufacturers will find that most of the inputs and services they require are available in Malaysia. In addition, production equipment is increasingly being manufactured within the country.

Industrialists wishing to establish contacts with business partners in Malaysia may obtain assistance and information from MIDA's Registry of Investors and Contract Manufacturers (RI-

COM). RICOM has a database of Malaysian companies which are seeking opportunities, either for joint-venture projects or for contract manufacturing.

7
Australia: Moving Towards Asia

PETER POND

THE dynamic economies of the Pacific Basin will provide the main economic driving force as we move into the twenty-first century and a new millennia and Australia is part of that region. The purpose of this chapter is to provide an overview of Australia, an attractive country that is wholly un-Asian, although situated at the edge of Asia, and to dispel some of the misconceptions which still exist about Australia.

Australia practices one of the world's most liberal and democratic political systems. It is known the world over for its golden beaches and beautiful surf; it has a magnificent Great Barrier Reef containing a myriad of fishes and other sea-life and providing the setting for many tourist destinations; it has miles upon miles of tropical rain forest; it has rolling temperate pasture land and produces wool, wheat, cotton and a vast cornucopia of fruits and vegetables; it has snow-covered mountains greater in land area than Switzerland; it has a unique admixture of flora and fauna, some of which are found nowhere else on earth; it is a major producer of minerals and metals such as gold, coal, iron ore, diamonds, bauxite, etc.; and it is a favourite playground of many of the world's rich and famous.

Australia has the highest GDP growth rate in the Organization for Economic Co-operation and Development (currently

3.3 per cent and forecast for 1994 at 4.0 per cent); it is the lowest taxing OECD country; in 1992 it had the lowest inflation rate in the OECD—1.0 per cent; its corporate taxation rate is an internationally competitive rate of 33 per cent; in the last two years its currency has fallen 20 per cent in trade-weighted terms and 30 per cent against the Japanese yen, making its products extremely competitive in world markets; over the past five years, manufacturing exports have increased faster than in any other OECD country (at an average annual rate of 15 per cent).

It has truly become a sophisticated manufacturing nation: sales of sophisticated manufactures to North Asia have increased by a remarkable 800 per cent over the past ten years while at the same time they have increased by more than 300 per cent to Southeast Asia; exports of sophisticated manufactures to the European Community have doubled since 1987-1988.

The Australian government has moved to make the country more competitive internationally: the local currency has been floated on world markets; the financial system has been deregulated; new competition has been introduced into the telecommunications system (it also has the highest usage of handphones per head of population in the world); air transport has been deregulated; productivity at ports (on the wharves) has doubled, and crewing rates on coastal shipping services have been reduced to OECD levels; its rail systems have been made more competitive and a vigorous programme for further improvement is well underway; progressive import tariff cuts are being made and by 1996 the general level of tariffs will be a maximum of 5 per cent; and there have been widespread labour market reforms.

Australia has truly made enormous changes in the past few years—in fact, some would go so far as to say it has truly "re-invented itself" in many ways. All these changes augurs well for the future of Australia.

The Changing Face of Australia

Australia has undergone pronounced changes in recent years. Its economy has internationalized—it has been transformed into an open, internationally competitive and growing economy. Of course, these changes have not been without a great deal of pain for Australia—and some of that pain is still continuing as the changes progress further, but they were, and remain, necessary if Australia is to be a major player in the regional and global markets in the next century.

I am sure many of you will be aware of—or as investors will have benefited from—the economic transformation to which I refer. But I suspect some of you may be a little sceptical about these changes in view of the outdated stereotypes so often presented of Australia and of Australians.

But let me give you a few examples. When people race a BMW, the gearbox is probably made in Australia. When anybody buys a ticket on Lufthansa in Frankfurt, they will do so with the aid of Australian technology; Australian software runs Lufthansa's airline trafficking, reservations and bookings systems. At some of the world's most advanced airports—Heathrow, Schiphol or Manchester—Australian software manages the airport system. If you travel on high-speed ferries in European waters, there is a very good chance that they were built by one of four Australian companies (in fact, it was only a few years ago that an Australian-built vessel set a new record for the crossing of the Atlantic Ocean. The Swiss Stock Exchange uses computer software developed by the Australian Stock Exchange. In 1987, an Australian-made satellite gave Vietnam direct telecommunication links with Australia and through Australia with the rest of the world—and today Australian-installed high capacity earth stations in Hanoi and Ho Chi Minh City provide Vietnam with direct international links everywhere. On public transport systems in Norway, Denmark, Sweden and the United Kingdom, the "smart card" ticket used by commuters is Australian-made. An Australian company will be providing the new container cranes for the Butterworth Port in Malaysia. A consortium of Malaysian, European and Australian companies is undertaking the electrification of the double-tracked railway lines in the Klang Valley around Kuala Lumpur. An Australian company will be building the carriages for Kuala Lumpur's new Light Rail Transit (LRT) system. Wilson Transformers, the largest Australian-owned manufacturer of power and distribution transformers based in Melbourne, provides power and distribution transformers to Tenaga Nasional Berhad in Peninsular Malaysia.

Market Reforms

The achievements mentioned above are a reflection of the changes which have been brought about in Australia. Scarcely an area of the economy has been left untouched by the reforms put in place by the Australian government. The financial markets were deregulated in the early 1980s—the floating of the Australian dollar and the relaxation of controls on the entry of foreign banks being two of the main initiatives. A modern legal system was introduced, featuring a unified corporation law with national coverage, to regulate unacceptable corporate practices and provide protection for investors. The taxation system has been substantially reformed and made more attractive for business investment—Australia's current corporate tax rate of 33 per cent being amongst the lowest in the OECD countries and comparative with those in the Asia-Pacific region.

In addition, there is now a range of specific investment incentives available from both Federal and State governments: a 10 per cent development allowance was introduced in 1992 applicable to certain large scale investments; an additional general investment allowance of 10 per cent for qualifying plant and equipment; the previous 150 per cent deduction for research and development expenditure has been continued; and accelerated depreciation arrangements on plant and equipment are in place enabling manufacturers to update their processing plant more rapidly.

Other reforms translate directly into lower cost structures. For example: the cost of domestic and international telephone calls has fallen by more than 20 per cent over the last two years (placing Australia well ahead of other countries in the region in terms of cost competitiveness); electricity prices have fallen by some 12 per cent over the last four years; air fares have fallen by up to 29 per cent since deregulation; and the real cost of labour has also continued to fall with the result that Australia is now significantly cheaper in terms of labour cost than other countries within and outside the region in many areas, particularly in the managerial and skilled areas.

Unions and the Labour Market

But what about the labour market? For many years, Australia's labour situation has been seen as a major disincentive for further investment. It once indulged unions to a point which sometimes went far beyond what was considered equitable in economic benefits or necessary in terms of of the power it gave unions to hold employers and the nation to ransom. This is no longer the case. Unions are slowly but surely changing their imperious ways. Industrial dispute is at its lowest ebb in forty years. More flexible and adaptable labour arrangements have been introduced on an enterprise-based wages system----enterprise bargaining---which enables employers and employees to negotiate wage agreements at enterprise level. There are now already well over a thousand Federal registered enterprise agreements in place covering about 13 per cent of the workforce---and this is expected to increase dramatically over the next few years.

Rather than the plethora of unions which managers had to contend with at a single plant or factory, there are more and more instances of single union shops. This is encouraging new investment. Toyota's new A$420-million motor vehicle plant built by John Holland in Melbourne is its most efficient operation outside Japan and the number of man-hours going into each car has dropped to around half the previous Australian levels.

On the waterfront, ship turnaround times have improved by 45 per cent—much to the annoyance of a Japanese ship captain who no longer has time for a game of golf while his ship is being unloaded and reloaded! Stevedoring and technical charges have been reduced by 20 to 25 per cent in the least year alone. The number of containers handled per employee shift on Australia's wharves rose by 127 per cent from 1989 to 1992.

Workplace reforms have meant major improvements in manufacturing. Kellogg's Australia, the breakfast cereal producer, has undertaken an A$400-million expansion programme as part of the implementation of its Asian Competitiveness Strategic Plan. Workplace agreements with the unions to support this large expenditure have meant that in 1991 there was only one industrial dispute at the plant and none in 1992.

The government has initiated reforms in occupational licensing leading to the more efficient use of skills in the workplace, the removal of demarcations, more efficient operation of plant and machinery and greater opportunities for workers to use their skills and abilities.

In the metals industry, 388 job classifications have been streamlined into fourteen categories of skills. Skills and remuneration have been better linked, and the introduction of skills-based classification structures have provided opportunities and incentives for employees to move either up or across the career structure.

Workers can now share responsibility with management for the competitiveness of their organization and, in return, share in the benefits that improved competitiveness brings. An example is the Sheraton Hotel agreement which provides for a reduction of staff classifications, annualized salaries in lieu of penalty rates and days off in lieu of overtime.

The education and training system now provides better incentives for training and retraining of employees aimed at achieving a more adaptable workforce.

These changes are leading to real improvements in productivity. Dulux Australia, at its Clayton plant in Victoria, have achieved greatly improved efficiency and productivity. The time taken to produce a batch of paint has been reduced from 120 hours to 40 hours---a very significant reduction indeed. Paint batches "produced in full and on time" and ready for dispatch improved from 68 per cent to nearly 100 per cent.

Technology

Australian research scientists and technologists are good at finding their own solutions when none is available elsewhere. Australia is now at the forefront of leading-edge research and development and is a recognized leader in a many research areas; its particular strengths lie in medical research, bioscience, engineering and manufacturing technology, environmental science and earth and space science.

The number of research scientists and engineers in research and development has increased two-fold in the past decade and employment in this sector as a proportion of the labour force has grown dramatically. Results of research include new packaging wraps to maintain the freshness in food, ways to keep livestock healthy, accurate measurement of coins, safer air travel, global climatic changes, etc.

Australian research scientists repeatedly demonstrate their capacity to combine creative thinking with the rigorous discipline needed for developing and testing new ideas, resulting in such advances as the Sarich orbital engine, gene shears for genetic engineering, Synroc for the stable storage of nuclear waste, the Interscan aircraft microwave landing system which is to be installed at all major airports in the world by 1998, the ready-mix system of transporting concrete, the black-box flight recorder used in aircrafts, the "balloon in a box" wine cask, the principle of photocopying, the atomic spectrophotometer, Sirofloc process to remove impurities and colour from water, bionic ears to help overcome nerve deafness from Cochlear lead the world market, ground penetration radar used for detecting mineral deposits deeper and with greater accuracy than before, lightweight plastic lenses for spectacles, in vitro fertilization, and heart pacemakers from Teletronics Pacing Systems capture 25 per cent of the world market. And the list goes on. According to the Paris-based OECD, Australia "has a sizeable scientific and technological potential and its research institutions show excellence in many areas."

Kenichi Ohmae, the respected Japanese industry consultant, strategic management guru and best-selling author of *The Borderless World*, offered his view of Australian industry on Japanese television when he said, "Australians . . . demonstrate the greatest creativity in the Asia-Pacific region in areas such as software development. This creativity can also be seen in the way they develop large areas of land. In the field of medicine and in their hospitals too Australians boast a level of advancement that is exceptional even by world standards."

Australia's Cultural Tradition

Lest it be said that these pronounced changes and developments have occurred in Australia at the cost of its culture, let me say that the tradition of cultural activity and achievement in Australia is as old as the cultures from which Australia's truly multicultural society evolved and as youthful as the nation itself. It is 60,000 years old----the reach of time of the aboriginal culture----and as young as *Strictly Ballroom*, the quintessential Australian film of youthful exuberance that took the Cannes Film Festival by storm back in 1992.

The Australian imagination soars high and unhindered in such wonderful films as *Breaker Morant*, *Gallipoli*, *'Crocodile' Dundee* and *Lightning Jack*. Australian performers and groups tour the world over to great acclaim. Australian rock music is a significant export-earner, and world-class Australian filmmakers, actors, musicians and arts administrators are working in Europe, North America and Asia. We have ballet, painting and literature.

Our most distinctive and celebrated building is probably the Sydney Opera House, and we have eight professional orchestras of substantial size. Australian sports has seen a resurgence in recent years with top performances in cricket, golf, squash, swimming, athletics, rugby, hockey, surfing,etc. Australians not only work hard---they play hard as well.

Trade Developments and New Directions

Australia has long been a leading trading nation and it still is. But there have been marked changes in both the composition and direction of Australia's exports over the recent years.

Australia is known as an exporter of primary products---the things we dig up and the things we grow. Most people will know of Australia's vast cornucopia of natural resources and as an exporter of wheat, wool, coal, iron ore, bauxite, lead, zinc, cotton, sugar, etc. Australia is the leading world exporter in many of these commodities, and in many others it is in the top five or ten exporters.

As you will have gathered from the earlier discussion about the changes in Australia, commodities are no longer as important as they once were. Australia is increasingly adding value to its raw materials before they are shipped overseas. In the past ten years, exports have increased from 13 per cent to 20 per cent of GDP and at the same time, the percentage of exports accounted for by manufactures has increased from 20 per cent to 29 per cent. Major manufactured exports now include office and computer equipment, electrical machinery, aircraft, passenger motor vehicles, telecommunications equipment, iron and steel products, medical and pharmaceutical products, worked aluminium and alloys, paints and varnishes, plastics, photographic supplies and equipment, textiles and copper products.

Australia has also become a significant exporter of services, including tourism and education; trade with Asean is now greater than that with the European Community.

Just as the composition of our exports has been changing, so has the direction of these exports. In 1957, when Australia signed its first trade agreement with Japan, the group of countries now comprising the European Community were taking 51.3 per cent of our exports while the countries of East Asia were taking a total of 21.2 per cent. That has now changed. Two-thirds of our trade is now with Asia. China will soon overtake the United States as our second-largest export market after Japan and South Korea is expected to displace the United States from third place by the turn of this century. In 1992, the proportion going to the European Community had dropped to 12 per cent and that to East Asia had grown to 58 per cent. Asean is now a larger market for our products than either the United States or the European Community.

The dominating factor in this turnaround is the growing realization among Australians that our future lies with Asia. There has been a corresponding realization among Asian countries that Australia, with the fourth-largest gross domestic product in the region, represents a reliable supplier of a range of goods and services that can fuel their own growth.

A recent survey of seven hundred emerging export companies showed that they were strongly focused on the Pacific Basin, particularly Southeast Asia. And offshore investment by Australian companies has also expanded rapidly---to A$106 billion in 1992-1993, representing 27 per cent of GDP, compared with 14 per cent in 1983-1984. This is to ensure that the Australian economy is locked into the overall global economy.

The internationalization of the Australian economy has also had its effect on the social and cultural aspects of Australian life. We now have more Japanese language students than any country, apart from Japan itself—over a hundred thousand, to be precise.

Australian Communications

Australia's telecommunications environment is one of the most liberal in the world. Reforms in regulations are leading to innovative services and underpin a rapidly expanding international demand for Australian products and services. And the winners are Australia-based corporations using the most up-to-date technology in their domestic and international operations.

Australia is ranked eighth in world market size for production of telecommunications equipment and the export of telecommunications equipment is the fastest growing sector of manufactured exports. Australian equipment manufacturers have won contracts for their innovative products in the United States, Indo-China and other parts of Southeast Asia. Exports to China, Singapore, Thailand and Malaysia are growing rapidly.

Alcatel Australia has built an A$100-million factory, the world's most modern cable manufacturing plant, to supply much of the submarine optical fibre cable for the Pacific Rim network. It is also contracted to supply cable for an optical link from Singapore to Europe.

Telstra, the umbrella organization for the export of Australian telecommunications services, expertise and equipment, is helping to establish centralized telecommunications and data processing in Australia to support business activities in the Asia-Pacific region. The benefits of a focused communications network centre, known as a hub, are in management control, economic efficiency, regional specialization and competitive edge. Hubbing is being used increasingly by corporations with multiple locations and a need for frequent internal communication by telephone and data transmission. Multinational corporations such as Barclays, County Natwest Bank and IBM are already taking advantage of Australia's place in the Asia-Pacific region and are using hubbing facilities.

Regional Headquarters

The Australian government recently announced a number of initiatives to increase business investment in Australia. In addition to securing a competitive tax system, the government established a National Investment Council to provide high-level policy advice to the Australian government on strategies to promote both Australian and overseas investment in Australia.

One of the first initiative undertaken under the auspices of the Council was the initiation of a campaign to encourage multinational companies to locate or relocate part or all their Asian regional headquarters operations in Australia.

This campaign epitomizes the watershed change in Australia's approach to attracting direct foreign investments and positioning itself in the international market and the region. It is an example of government and industry working together hand in hand to capture new markets and attract productive investments into Australia. For companies located outside this region, the establishment of regional headquarters in Australia makes good economic sense.

Other Investment Opportunities

There are many other investment opportunities in Australia. Opportunities exist in iron and steel, petrochemicals, petroleum, natural gas, non-ferrous metals and other minerals, pulp and paper, wool processing, skins and hides, food processing, information technology and telecommunications.

Conclusion

These are some of the excitement I feel with regard to the changes that are happening in Australia. These changes are not occurring at the cost of all those things which we appreciate about Australia. It is still very much a cosmopolitan country with a huge range of environments—from the concrete canyons of the inner cities to the arid wastes of the Great Australian Outback; from the hot and humid tropics to the snow-capped mountains of Tasmania. And you can still play golf there at unbelievably low prices.

Australians are still warm-hearted and friendly; they are amongst the most egalitarian people in the world, growing daily in their understanding of their neighbours and their desire to seek opportunities for mutual development. Malaysians are now visiting Australia in increasing numbers to see for themselves---both the "old" Australia, with its magnificent tourist attractions, and the new, "re-invented" Australia with its high-technology, value-adding investment opportunities.

8

Collaboration for Success in the Pacific Rim Decade

PETER COOK

IN this age of globalization and as we enter the Pacific Rim century, there is an increasing need for regional businesspeople to be able to tap into timely business data and develop contacts with their business counterparts in other countries. It is also important for people to be well-trained to meet the rapidly changing demands of the business world.

Australia's important role within the Pacific Rim and its commitment to business collaboration within the region cannot be over-emphasized: Australia and Malaysia have much to gain from working together, from working with and not against each other.

The theme of this chapter is the underpinning of business collaboration: shared interests. About a decade ago, two members of the Harvard Negotiation Project wrote a book called *Getting to Yes*. It has been acclaimed by business and political leaders throughout the world as a first-rate guide to negotiation and identifies a number of principles designed to lead to win-win outcomes. A key principle is to focus on interests, not positions. Interests, not positions, is what we should all be about. That is what successful business leaders do when they negotiate their business deals. And that is what governments must also do.

In discussing shared interests, there is no better place to start than with the Uruguay Round and its relevance to the Asia-Pacific region. With the demise of the Cold War and the fading away of ideological disputes, hundreds of millions of people have entered the global trading system. If the new world order is about anything, it is about the importance of economic growth, increased trade and investment flows, and the emergence of market economies.

We are now at a critical juncture in world affairs. And we are poised on the threshold of a truly global and universal trading system. The latest news from Brussels, Geneva and Washington, D.C. suggests that the United States and the European Union are close to settlement on agricultural and industrial issues. The services package appears to have a little further to go. We need to see the fine print before we can judge the quality of these outcomes. And, of course, none of them can be signed off by the rest of us until satisfactory market access packages are concluded.

The clock is fast ticking down. We are now at the point where we should all press for the biggest market access outcome. We should push the majors harder, and objectively review our own offers to provide the momentum for the boldest possible outcome.

The countries in this region currently enjoy the highest levels of trade growth in the world and stand to gain a great deal from a positive outcome of the Uruguay Round. Ironically, some of the countries in this region have put forward some of the most conservative market access offers, but I am confident that the final negotiating phase will see improvements in this position.

Australia shares a deep interest in a more open global trading environment and has made clear its willingness to contribute its share to achieving this important objective. For Malaysia, with its high export orientation, an open global trading environment is no less important.

Australia and Malaysia share many common interests in many different sectors in the Uruguay Round. Both of us are increasingly exporters of industrial goods and services. But we also retain a strong interest in improving the conditions under which most commodities are traded.

This is why Malaysia and Australia have worked so closely together in the Cairns Group for a positive outcome on agriculture. Malaysia with its important palm oil industry can only benefit from a less distorted, more liberal oil and oilseeds market. Malaysia also stands to benefit directly from the outcomes on tropical products.

While membership of the Cairns Group is based on shared interests rather than any criteria of geography, historical linkages or levels of economic development, it is a fact that ten of its fourteen members are countries of the Pacific Rim.

If we—and I expect we will—get a balanced result on agriculture, then this region will benefit significantly from it, and the Cairns Group members, including Malaysia and Australia, will truly be said to have collaborated for success.

Let me now turn from the changes in the world trading environment to examine the economic developments which have been occurring in the region. In the 1980s, Northeast Asia was the fastest-growing area in the world with real GNP growth averaging more than 7 per cent—a doubling every ten years. In 1980, the economies of the region accounted for 41 per cent of the world's GDP; today they account for more than half. The region's growth continues to outstrip world economic growth.

Much is owed to the vision of strong national leaders in the region, and to their purposeful commitment to economic growth. High rates of savings and investment, high priority to improved education and sound macroeconomic policies have provided the essential conditions for some remarkable achievements.

The rapid growth has been centred on export-oriented industrialization involving increasing international competitiveness. Throughout the region protectionism has been significantly reduced—though more remains to be done, foreign investment has been welcomed, and dynamic export-oriented economies have flowed from outward-looking policies. This formula is proven.

Open markets have been the lifeblood of highly export-oriented economies. The results are there to be seen in the trade figures. In 1980, exports from the Asia-Pacific economies were valued at US$593 billion and accounted for only 31 per cent of total world exports. In 1992, they had leapt to US$1.5 trillion, and accounted for 41 per cent of the world's exports. The remarkable thing is not just the size of this growth, but the degree of integration that has come with it, reflecting the extensive complementarities and linkages amongst the Asia-Pacific economies.

The percentage of intra-regional trade is rising fast. Today, two-thirds of the trade that the Asia-Pacific countries do with the world, they do with one another. Intra-regional imports are just over two-thirds of total imports and growing—the Asia-Pacific economies are increasingly sourcing their imports from other Asia-Pacific economies.

Similarly, on the investment front, while it is difficult to analyse the flows within the region because of statistical inconsistencies, we do know that intra-regional investment flows have increased dramatically. We know, for instance, that 60 per cent of the foreign direct investment that has flowed into Southeast Asia from 1987 to 1990 came from other economies in East Asia.

Japan has been the main source of this intra-regional investment, but investment flows from South Korea and Taiwan into China and Southeast Asia and from Southeast Asia into China and Indo-China have been increasingly important. Over the past decade, Australia has increasingly been a key player in the regional investment scene. The result is that the Asia-Pacific region already enjoys a higher degree of economic integration than the countries of the European Union. This has largely been due to business collaborations driven by market forces and appropriate government policies.

Is it any wonder then that the countries of the region have such a strong, shared interest in open regionalism. But this vision is not shared universally. We know from experience that the region has markets closed to it—and that the trading environment frequently is not free or fair. Protectionism will not end just because the Uruguay Round succeeds.

Strong voices are raised from time to time in both North America and Europe arguing that the rapid and successful growth in Asia is a threat to the economic and social fabric of those parts of the world. This is a new form of protectionistic argument. Unfortunately, it is also often manipulated for narrow political reasons. But the truth is that this region of the world represents incredible business opportunities which are beneficial to all.

Both Malaysia and Australia have contributed to these growth trends in the Asia-Pacific region, and both are in turn benefiting from them. Malaysia, on its part, has recorded one of the highest growth rates in the world for over a decade. It offers a liberal trading regime and a coherent industrialization strategy aimed at developing a competitive export-oriented manufacturing base with an increasing share of high value-added manufactures. The Malaysian government's strategy is to develop capital- and technology-intensive industries by offering generous and attractive incentives---liberal equity participation rules, employment of expatriates, pioneer status, investment tax allowance, reinvestment allowance, etc.---to lure manufacturers, especially those investing with a view to exporting from Malaysia.

Malaysia has thus positioned itself well to attract investments and to benefit from an access to growing markets in the region. As a result, more than 76 per cent of all Malaysia's exports go to other Asia-Pacific economies. While Japan remains the biggest foreign investor in Malaysia, the Australian presence is getting stronger. Australians are impressed with Malaysia's strong economic performance and future potential. In 1992, Australia was ranked fourth in the foreign investment stakes behind Japan, the United States and France, up from ninth the previous year. In 1993, Australian investment in Malaysia surged to US$1.25 billion from US$243.5 million in 1991.

The Australian story is similar. In the mid-1980s, we made a conscious decision to open our economy to the world. And the benefits of the structural change are now showing strongly in the composition and direction of our trade.

Today, exports of manufactures now exceed exports of rural products and minerals and fuels. Manufactured exports increased by an average of 12.5 per cent per annum between 1982 and 1992—more than four times faster than growth in agricultural exports and slightly faster than growth in mining exports. Southeast Asia provides an important and rapidly growing market for our manufactured goods.

In the faster-growing component of world trade, namely, services, our exports have grown threefold since 1983 and now account for almost a quarter of our total exports. And our trade, like Malaysia's, is heavily focused in the Asia-Pacific region. Asia-Pacific economies take about three-quarters of all Australian exports and account for twelve of our thirteen largest markets. And two-thirds of our imports are sourced from the Asia-Pacific economies.

Sixty per cent of Australia's export trade is with East Asia compared with 46 per cent ten years ago. Forecasts suggest that eight out of Australia's top twelve export markets between 1994 and 2012 will be in East Asia.

Given this convergence of interests, it is not surprising then that Australian and Malaysian companies, having long worked together in productive joint ventures, are now increasingly seeking opportunities for business alliances in third countries.

Typical of these partnerships in the region has been the alliance between Australian and Malaysian companies to develop the Dai Hung oil field off the coast of Vietnam.[6] This is an investment worth around US$1.5 billion. BHP Petroleum of Australia is the leader and Petronas, the Malaysian state oil company, is a

---

[6] The offshore Dai Hung oil field is Vietnam's largest oil field. Dai Hung is about 250km southeast of the coastal town of Vung Tau.

major stakeholder in the consortium which beat eight other international competitors to win the bid to develop the Vietnamese oil field.

Another such alliance is Slumberland bedding, which has long been a market leader in Malaysia. The success of this co-operative endeavour between Australian and Malaysian entrepreneurs is now paying dividends from a joint venture in the fast-growing Chinese bedding market.

For our part, we applaud the circumstances which have given rise to these alliances and collaborations and are committed to maintaining a healthy environment for the expansion of business alliances and collaborations between these two countries. We welcome particularly the fact that the benefits that follow from our joint ventures and business partnerships flow to both countries.

By working together and by focusing on shared interests, all countries in the Asia-Pacific region have secured benefits, and the benefits have been mutual. We all want to see continued high growth rates and improved living standards. We all want to ensure that we retain access to regional and global markets and where possible expand that access. And we all accept that the best way to ensure this is to work together to guarantee stability and security for those markets.

It is in this context of working together for the future that Australia attaches such importance to APEC,[7] the Asia-Pacific Economic Co-operation. APEC has developed in the way it has because of a growing consciousness of shared interests in open regionalism. APEC members of course have a shared commitment to a successful conclusion to the Uruguay Round of GATT. In

---

[7] APEC is the four-year-old organization of seventeen economies: the six Asean member countries (Brunei, Indonesia, Malaysia, the Philippines, Singapore and Thailand), Australia, Canada, China, Hong Kong, Japan, Mexico, New Zealand, Papua New Guinea, South Korea, Taiwan and the United States.

Seattle, Washington, APEC ministers called for a settlement package covering US$250 billion of trade in specific sectors. But the ministerial meeting in Seattle went further than just calling on others to act. Ministers set their sights squarely on making it easier to trade and invest in the Asia-Pacific region. Work is proceeding apace on such matters as:

1. The harmonization of standards and testing and certification arrangements;
2. The recognition of professional qualifications;
3. The improvement of customs procedures, including the promotion of paperless trade through the electronic data interchange; and
4. Greater transparency in existing investment regimes and the identification of impediments to investment.

As we see it then, APEC helps assures a more secure and open global and regional business environment. In pursuing this goal, APEC has not forgotten its origins: it will not become an institutional, bureaucratized international agency. It will not compete with governments. Rather, it will look for business.

Our aim, in large part, was to contribute to the development of Australia's export culture. But we achieved more than this—real business was done, and new networks were established. But we do not of course see APEC as the only forum for pursuing this objective. Bodies like the Pacific Economic Co-operation Conference and the Pacific Basic Economic Council, among others, and a host of think-tanks around the Pacific Rim, are working towards the very same goal.

I have dwelt on the collaborative exploitation of shared interests on a number of different planes—global, regional and bilateral. Throughout this discussion there has been a consistent reference point—the spectacular economic success of this region. The fact is that we are participating in a more fundamental change in the shape of the global economy than at any time since the

Americas emerged as a significant alternative centre of economic activity to Europe in the nineteenth century.

The pace of that change is accelerating and I am confident that it will continue to accelerate. And by the time Dr Mahathir Mohamad's Vision for Malaysia in 2020 is achieved, the pattern of world trade and wealth distribution will be even more different from how it was in 1945, when 40 per cent of the world's GDP was generated in the United States. But we already know this. Of greater interest are the efforts made by many in our region to build for that future which we know is coming. And what we need to build is the best possible economic environment in which our shared interests can be exploited further for mutual benefit in the years to come.

For Australia, the imperatives are clear—an open global and regional trading environment, strong, outward-looking regional economic organizations and significantly greater economic engagement with the region in which we live. For the region as a whole, the way forward for us all ought to be equally clear. We need to utilize growth and change in our region to achieve a global economic order based more on mutual economic interest and less on insular protectionist sentiment than it has been for the whole of this century.

The tectonic shift in the world economy towards our region presents us with many challenges and opportunities. Opportunities exist for us to collaboratively create a global economy led by a successful Asia-Pacific towards a more open, liberal and prosperous world.

9

Canada's Trade and Investment Interests in the Pacific Rim

JOHN DAVID MULRONEY

DESPITE the ambitious and vast geographical sweep of this chapter's title, I will only discuss Canada's trade and investment interests and opportunities in Malaysia. At times, we have to work real hard to convince people that we do indeed have interests and opportunities here in the region. Let me illustrate with a quote from a recent interview with the Malaysian International Trade and Industry Minister, Rafidah Aziz, in the *The Financial Post*, a Canadian business paper: "Canadian entrepreneurs should be thinking globally---beyond NAFTA. Perhaps a large country is always a bit smug and insular. Maybe people who live in a cold climate tend to cocoon [themselves] and be more inward looking. They're afraid to venture out in case it's even colder outside."

Of course, one of the great things about good friends is that they can speak frankly to one another. We Canadians appreciate those exhortations to do more and to try harder here. But you must excuse us for our typical polite bewilderment and frustration that many of the good things that we are already doing here and in the region tend to go unnoticed and unrecognized. There are of course some obvious reasons for this: physically, we are too far away; we are known mainly as a tourist destination and for our vast cornucopia of natural resources and not for our knowledge-based industries; and we allow ourselves to be overshadowed by our self-confident neighbours to the south.

The main thing is that we Canadians are by nature reserved and not given to blowing our own horns. Everybody knows that Canada is the second largest country in the world with a total land area of about 10 million square kilometres, stretching from the Atlantic to the Pacific; indeed a big country with a small population. Though they somehow tend to forget that while we are only thirty-first in terms of population, we have the world's eighth largest economy and is one of the G-7 countries. According to the Organization of Economic Co-operation and Development (OECD), Canada will have the fastest growing GDP among the G-7 countries in 1994. We also have the lowest inflation rate[8] and are projected to have the highest growth rate in 1994. We are here talking about a 3.5 per cent range----not spectacular in Asian terms---but not bad for a gross domestic product of close to 700 billion Canadian dollars. We are also expected to lead the G-7 in employment growth in 1994. Looks like we are entering a phase of renewed economic dynamism.

Canada's economic strength lies in its diversity. Yes, we are the world's leader in the export of wide a range of natural resources,[9] but that is far from all we do. You may not know that more people in Canada work in the electronics industry than in the pulp and paper industry. We have an automobile industry that is measured in the tens of billions of dollars each year but our computer services sector employs more people. And in British Columbia, the source of much of the Pacific Rim's softwood

---

[8] Canada's inflation rate of just 1.5 per cent in 1992 was a post-war low.

[9] Canada has evolved from an economy dependent on resource-based activities such as petroleum and natural gas extraction and refining, wheat, minerals and lumber, to one of technology-based such as services, manufacturing, transportation equipment, electronics, robotics, computers and telecommunications.

lumber, more people are employed in the communications industry than in forest products. In fact, it is precisely the challenge of efficiently extracting, processing and exporting our natural resources that has enabled us to develop leadership in areas like transportation, communications, financial services, insurance, real estate, power generation, oil and gas technology and environmental protection technology. Companies in these sectors are backed by and often spun off from a science and technology base that is second to none. The Nobel Prize for Chemistry in 1993 happens to be shared by Professor Michael Smith of the University of British Columbia and we had the honour of hosting another distinguished Canadian Nobel Laureate, John Polanyi, in Kuala Lumpur a few months back.

And let no one underestimate the importance of international trade to Canada; Canada is a trading nation. In 1992, Canada's merchandise exports accounted for C$163 billion. Imports stood at C$148 billion for a surplus of C$15 billion on the merchandise account. So far this year Canadian exports are up by 15 per cent. When you consider that every billion dollars in exports represents some fifteen thousand jobs, you can see that all Canadians have a stake in promoting international trade. In fact, a quarter of all Canadian jobs and a quarter of our national wealth depends on trade. That's a huge investment in the global economy.

It should come as no surprise to many people that much of Canada's trade traffic flows on a North-South axis and that the United States is our single most important partner. Now, you can look at that from many perspectives, but I think you will agree that the benefits of familiarity and proximity, plus the dismantling of tariff barriers through the North America Free Trade Agreement[10] place Canada in an enviable position. In 1992, that

---

[10] NAFTA reduces trade barriers within North America (Canada, Mexico and the United States), but allows each of the three countries to maintain its existing trade regimes with other trading partners.

meant a two-way trade of C$220 billion out of which Canada enjoyed a surplus close to C$30 billion.

Perhaps because it is so smooth and relatively problem-free, people tend to forget that Canada and the United States have the world's largest two-way trading relationship. It's a mistake, though, to think that we have all our eggs in one basket. In 1992, we did about C$88 billion worth of two-way business with the rest of the world. That includes about C$18 billion with Japan and close to C$5 billion with China. In fact, Asia is the destination for close to half of our non-US exports. That provides an obvious answer to an obvious question: whether Canada's trade interests extend beyond the huge North American marketplace.

Of course, you don't just have to look at the statistics. Take a walk through downtown Vancouver or check the itineraries of businesspeople from Calgary, Toronto and Montreal. The Asia-Pacific region represents an increasingly important destination for them. Flows of trade, investment and people to and from the region are surging. Canada's trade across the Pacific surpassed that of our Atlantic trade a decade ago, and the gap is widening.

As heartening as these developments are, particularly for those of us who are committed to expanding our trans-Pacific trade, let there be no illusions that we in Canada are motivated by anything other than enlightened self-interest. We recognize where the opportunities are and also understand that global success means competing with the best in the world's most dynamic markets. We know that the Pacific Rim has emerged as the most vibrant region of the world and there are signs that world economic activity is gravitating from the Atlantic to the Pacific.

What's more, there is a new and more sophisticated focus on opportunities within Asia itself. Recently, the Conference Board of Canada concluded a study of five hundred Canadian firms with established credentials in Asian trade, asking them to rank priority markets. Asean led the list, ahead of such huge and long-established markets as Japan and China.

Interesting, too, was the response to a question about business interests: while 55 per cent of the respondents said that they were involved in the shipment of goods to the region, the remainder characterized their interests as everything from technology transfer to licensing agreements to various forms of equity and non-equity joint ventures.

A quick review of the Malaysia-Canada trade relationship helps shed light on how Canadian entrepreneurs are transforming those interests into reality---and how the Canadian government is supporting not just traditional export trade, but the range of activities captured under terms like industrial co-operation and technology transfer. In summing up that relationship we tend to focus quite naturally on the export of goods, something that is readily measurable. The news here is good for Malaysia, with your exports up 41 per cent (to more than C$400 million at mid-1993), though not so good for us, with exports to Malaysia down 25 per cent.

But there is another reality that is much harder to measure. Looking closer at Canadian exports to Malaysia, you will see that they are dominated by commodities which are down this year, and that processed and manufactured products are, on the contrary, growing steadily. Furthermore, what doesn't get captured in the trade statistics are exports of Canadian services in areas like designing of pipelines, banking and finance, and the planning of transportation systems. And Canadians are also putting their money where their business interests are. Investment players range from well-known Canadian-based multinationals like Alcom, Alcan, Nova, Bata, Dominion Textiles, Bell Canada International and Northern Telecom to smaller firms specializing in areas like speciality chemicals, hardware products and ice hockey equipment.

The Canadian government on its part recognizes that this process of globalization is essential to maintain the viability and competitiveness of our companies. Indeed, it was to ensure this outward orientation and readiness to take on the world that we embraced the Canada-United States free trade agreement and are leading proponents of regional and global trade liberalization.

But we are doing more than cheering from the sidelines. New programmes have been introduced to help Canadian companies increase their staying power in new markets. Trade promotion funds have been reoriented to support long-term marketing plans with plenty of scope for return visits. In 1990, an industrial linkage programme was established by the Canadian International Development Agency (CIDA) to foster closer economic co-operation between Canada and Malaysia. Enterprise Malaysia Canada (EMC) is a C$15-million private-sector driven programme funded by the Canadian government that brings together prospective joint-venture partners from Canada and the region. We have set similar programmes in other Asian countries such as Thailand. It has offices in Kuala Lumpur, Vancouver, Toronto, Montreal, Calgary, Edmonton, Winnipeg, Ottawa, Quebec and Halifax that are working full time to identify and promote opportunities for Canadian firms to invest and joint venture in Malaysia.

Enterprise Malaysia Canada champions Canadian private sector investment in Malaysia. It sources the best in technology and know-how from across Canada to suit Malaysian investors' project requirements or offer them new investment opportunities. It offers a comprehensive database of opportunities, knowledgeable and experienced advisers in both countries, close liaison with key government and private sector contacts in Canada, the ability to match investors' needs with Canadian partners and provide financial assistance or grants to eligible Malaysian companies for exploratory visits, viability studies and other pre-joint venture studies.

The main purpose of the programme is to assist Malaysia in acquiring the necessary technology and expertise in its industrialization drive through joint ventures and other mutually beneficial commercial relationships such as strategic alliances and licensing agreements.

The Canada-Asean centre, based in Singapore and working on the premise that Canadians have to start "hunting in packs", has begun encouraging the formation of business networks. These bring together like-minded entrepreneurs in carefully focused consortia. The environment and rural telecommunications are two current areas of interest and emphasis.

I have focused my discussion on Malaysia and Asean, but Canadians are at work throughout the region. From Tokyo, where our new embassy is a showcase not just for Canadian architecture, but for a steady stream of Canadian products and technologies; to South Korea, where some of the longest lines at this year's Taejon Expo were in front of the Canadian futuristic Pavilion; or Hong Kong, home to tens of thousands of Canadians and Canadian graduates. On its part, the new Canadian government has its sights set on the Asia-Pacific region. While every effort has been made to streamline government structures—and this is most notable in our new, smaller cabinet—we have also created an important new office, that of Secretary of State for Asia. And the first person to hold this office is the Honourable Raymond Chan, a Member of Parliament from Vancouver.

I have tried to make a case for the fact that Canada's role in the Pacific Rim has to do with more than geography and that our Pacific Boundary has indeed become a front door and that we Canadians have a clear idea of where our future lies.

10

Mexico: Strategic Alliances for Prosperity

JOSE LANDA

BEFORE Mexico liberalized its economy eight years ago, it suffered huge deficits and exceptionally high inflation rates—as high as 159.2 per cent in 1987. And as Mexico liberalized its economy, inflation fell and deficits were reduced. For the past eight years, Mexico has undertaken major economic reforms to liberalize trade and to attract direct foreign investment.

This chapter highlights Mexico's rapidly changing business climate—a climate which is now amongst the most pro-business in the world, and one that actively encourages and welcomes direct foreign investment. In this chapter, we will explore the macroeconomic changes that have taken place in Mexico: the rapidly growing attractiveness of Mexico's domestic market, operating costs and productivity, cases of foreign companies that have been highly successful in Mexico, an overview of the regulatory climate for investing and operating in Mexico, and areas for investment and trade opportunities.

Mexico's macroeconomic policies have resulted in tremendous stability, fewer trade restrictions, an active stock market, a rapid privatisation programme, and a great deal of liberalization in the regulations governing direct foreign investment. Mexico is actively selling itself as an investment gateway and as an avenue to other Latin American countries.

Mexico is strategically located between the huge North American markets of the United States and Canada and South America, and has good access to Europe and Asia. Our trade with the outside world has grown dramatically: total exports and imports each exceeded US$46 billion in 1992. Despite the fact that the United States accounted for well over half of these trade flows, our trade with Europe and Japan is significant and continues to grow.

Companies invest in Mexico for many reasons: as a base for serving global markets; to serve the local market of 85 million consumers; and, of course, to reduce operating costs. But the fundamental reason is that Mexico is a good place to do business. Interviews conducted by A.T. Kearney, a management consulting firm in the United States, with executives of corporations in North America, Europe and Japan who have operations in Mexico revealed that their Mexican operations are frequently amongst their most profitable foreign ones.

The reason for Mexico's change has been a political leadership that embraces private enterprise, foreign investment, internationalism and government-business co-operation. The end-result of all the reforms has seen a dramatic increase in business confidence. First, Mexico's risk rating has improved significantly over the past five years. We were regarded as one of the higher-risk countries in the early 1980s, but now we rank as one of the lowest-risk countries of the world.

The success of Mexico's economic adjustment programme has given the country widespread international recognition. An example of this is the recent proposition that Mexico become a member of the Organization of Economic Co-operation and Development.

Based on these positive changes, Mexico's risk rating is improving rapidly. Not only has the paper issued by Mexican entities been upgraded due to an overall improved economic performance, but the country as a whole has enjoyed improved risk rating.

Another important indicator of international acceptance of Mexico's economic performance is the prevailing stability of the currency exchange market. Exemplifying this is the fact that the official reserves of the central bank of Mexico are at their highest level at over US$23 billion.

Foreign debt as a percentage of GDP has fallen dramatically since 1986. In 1986 and 1987 foreign debt accounted for 90 to 95 per cent of GDP. As most of you know, Mexico experienced a foreign debt crisis in the early 1980s. This has largely been solved, and consequently our foreign debt has fallen from over 80 per cent of the GDP in 1986 to about 20 per cent in 1992.

Mexico compares favourably with other low-debt countries. The strict control over public sector external and domestic borrowing has drastically reduced the proportion of the gross public debt as a percentage of GDP from 68 per cent in 1988 to 22 per cent in 1993. This control has allowed the government to allocate more resources to social spending, since the debt servicing to GDP ratio during this period of time has decreased from 17.5 per cent to 3.0 per cent.

The Mexican government has increased its control over public expenditure. The public deficit as a percentage of GDP has fallen from approximately 15 per cent five years ago to a one-percentage surplus today. Tight fiscal controls and a policy of encouraging privatisation have been major contributors to this. For example, the government strongly encourages private investment in infrastructure projects such as major highways.

Another improvement has been the reduction in the public deficit which was close to being in balance in 1991 and shows a small surplus from 1992.

The relevance of this well-balanced budget to the country lies in the urgent need to finance social expenditure in a non-inflationary way. With strict measures undertaken, social expenditure has been growing slowly to reach 9.5 per cent of GDP in the 1993 fiscal year.

Although Mexico is not affected by the Maastricht criteria, our efforts to attain and sustain the country at international levels is well illustrated by this comparison. This criteria defines a well-balanced budget as one which maintains a deficit of less than 3 per cent of GDP (or a surplus) and a gross public debt less than 60 per cent of GDP.

Another corner-stone of building business confidence in Mexico has been the huge reduction in our rate of inflation. According to the Bank of Mexico, our inflation rate too has plummeted; Mexico's single-digit inflation rate today is similar to that of other leading global countries. Mexico has been a leader in bringing inflation under control. In 1987, the inflation rate was 159.2 per cent, 29.9 per cent in 1990, 18.8 per cent in 1991, 11.9 per cent in 1992 and 8.0 per cent in 1993----within striking distance of those of the major world economies with whom we trade. This has been achieved through the economic and fiscal discipline of the government of President Carlos Salinas, and in particular through the truly remarkable social pact which ensures co-operation amongst the government, businessmen, peasants and unions. Salinas' free market reforms, cuts in the budget deficit and restructuring of the foreign debt have done much to open Mexico's economy to the world and win approval from international bankers and financiers. And the inflation rate for 1994 is expected to be around 5.0 per cent.

Tariffs have been reduced dramatically since the early 1980s to encourage foreign trade. For example, in 1982 the average import tariff was 100 per cent. Today the average import tariff is only 10 per cent, with a maximum rate of 20 per cent. Regarding restrictions, only 1.7 per cent of goods entering Mexico today require licenses, against 100 per cent in 1982.

We will now discuss trade----beginning with export growth. Mexico's increased status as a world trading partner is evident by its growing exports which have doubled since 1986. Not only have they expanded, but they now include a wide range of manufactured goods. These include manufactured products, for example, vehicles, chemicals, electronics and natural resources, ranging from oil and minerals to agricultural products. Mexico exports to virtually every major economy of the world indicating that increasingly our products meet the world's quality standards.

With oil declining and manufactured goods increasing, the composition of Mexico's exports has changed dramatically. In 1982, oil accounted for 80 per cent of exports, but by 1992 this had dropped to 52 per cent as the Mexican economy diversified.

Mexico is also an active consumer of foreign goods. Imports had quadrupled over the past six years. Liberalization of import restrictions and the reduction in tariffs to an overall average of about 10 per cent stimulated trade. It should be mentioned, though, that the rate of growth of imports has shown an important decrease in the first half of 1993. Whereas imports last year grew at a rate of 35 per cent, this growth in 1993 fell to 7.5 per cent. This is largely the result of new foreign investments which are now supplying the domestic market from their newly established facilities in Mexico.

Mexico has opened the door to its markets. There has been a dramatic change in the business climate in Mexico over the recent years. No longer is the economy protected from global competition. There is easier market access, more privatisation, more favourable foreign ownership regulations, and a growing amount of stock market activity which is a vote of confidence for these companies.

It now has a liberalized investment regime. Seventy-five per cent of the economy is now open to 100 per cent foreign investment. The major sectors which remain closed to foreign ownership include the energy and transportation industries (such as the country's airlines) and certain parts of the chemical industry. However, the government believes that in order to encourage even more foreign investment, new and more flexible regulations are needed. Therefore, a new, more liberal foreign investment law will soon be presented before Congress. Foreigners are encouraged to invest in sectors such as telecommunications, infrastructure and tourism.

In 1991, cumulative direct foreign investment in Mexico stood at US$40 billion. It now surpasses US$50 billion, which is almost three times what it was in 1985.

The growing confidence in the achievements of the Mexican economy is reflected in the country's ability to successfully attract foreign direct investment: Mexico ranked eighth in 1991 of all the so-called emerging economies of the world.

The United States is Mexico's largest foreign investor. Sixty-two per cent of Mexico's foreign investment comes from the United States, followed by 26 per cent from Europe and the balance from Asia and other countries. The largest investors outside the United States are the Germans, the British and the Japanese.

Most of the investment is in the industrial sector. Almost 56 per cent of foreign direct investment concentrates in industrial projects, whereas 32 per cent goes to services.

There is a growing list of companies with significant investment interests in Mexico---tangible evidence of the upsurge in business confidence in Mexico. Our foreign investors number in the thousands, and include most of the world's most renowned global companies. Many of these companies are planning further investments, including upgradings and expansions to their existing operations.

Mexican Market

I would now like to turn to one of the major drivers of direct foreign investment in Mexico: our large, growing market. Mexico has become a very attractive market for a wide range of manufacturers and suppliers. Its young population with their demand for sophisticated goods and services and the overall growth in the economy now makes Mexico the thirteenth largest economy in the world. Furthermore, it is a consumer-driven economy, like its neighbour to the North.

Mexico is a newly industrializing country. With a booming population of 85 million people, it has a large domestic market. Mexico also has the tenth largest potential consumer base in the world. It has a large population of 85 million people which is considerably more than South Korea and Taiwan in East Asia and exceeded only by Brazil in South America.

The bulk of Mexico's population is in an age group of active consumers. Not only does Mexico have a large population, 85 per cent of this population is 40 years old or less. In other words, it has a large number of young people approaching their most active earning and spending years.

Its increasing GDP per capita has lead to a growth in purchasing power. Mexico's GDP is growing more rapidly than its population. This not only implies a significant increase in real per capita GDP growth, but is also an indicator of the real increase in the people's future purchasing power. The effects of the increase in real income and a stable price system are reflected in the demand for consumer goods.

And as Mexico's economy grows, market opportunities will continue to grow in the auto industry. This is only a sectoral example of the potential of Mexico's domestic market: the average age of an automobile in Mexico is 8.65 years. This, together with the increase in real income, is expected to have both a direct and indirect effect in the demand of the overall market.

Major OEMs continue to invest in Mexico. It is estimated that by the year 2000 Mexico will be the fourth largest manufacturer of vehicles worldwide.

Suppliers of automotive parts are thriving. Consider the indirect effects of this industry on economic activity. But this is only one of several examples.

Opportunities for more penetration of consumer electronics. The consumer electronics market in Mexico is another attractive and rapidly growing one. For example, sales of video-cassette recorders are just beginning to take off, and the estimated penetration figures show that there is enormous potential for future growth in the coming years.

Spending on food is increasing. Processed food is another area of rapid growth. Just as in Europe and the United States, the pressures of two-career families with their lack of time for cooking are stimulating demand for processed foods. This is reinforced by the growth in the middle class and more sophisticated tastes. And demand for healthcare products is growing as the Mexican economy improves.

From a business point of view, in the years to come, Mexico will be more stable, economically and socially. Mexico is truly a new world for business: its economy is improving rapidly and government regulations have been reduced substantially, thus enabling businesses to operate in a largely open and competitive environment. There are significant demands unmet making the Mexican domestic market attractive for global manufacturers. This, in turn, is linked to the increased purchasing power of Mexico's population, particularly taking into account the growing affluence of its middle- and upper-income classes.

I would like to discuss a little bit about operating expenditure in Mexico beginning with some comments on labour cost and productivity. Mexico has one of the most cost efficient labour force in the world. The data published by the World Economic Forum in its "World Competitiveness Report" illustrates how Mexico measures up in terms of unit labour costs, and by that I mean labour rates and productivity. First, let us take a look at the labour rates (including fringe benefits). As you can see, Mexico's labour rate of US$2.17 per hour is far lower than that of the United States, Germany or Japan. It is also less than the rates in Taiwan, South Korea or Brazil.

Its productivity compares favourably with that of other developing countries. It takes a little over eleven workers in South Korea and fourteen in Brazil to produce what ten workers in Mexico produce. Mexico came out ahead in the index of total labour costs. Putting the two together, that is labour rates and productivity, Mexico came out ahead of countries like Brazil, Taiwan, Japan, South Korea, Spain, United States and Germany. For example, unit labour costs are approximately four times higher in Germany than in Mexico and about three times higher in the United States than in Mexico.

This comparative advantage will prevail. In September 1993, an initiative was taken to link real wage increases to productivity. This was implemented as a non-inflationary measure designed to create real wage increases, fuelling the growth in domestic demand.

Let us take a look at the cost comparisons of pharmaceuticals. This cost comparison applies to the manufacturing of injectable drugs. Overall costs are estimated to be 25 per cent lower in Mexico, due to much lower overhead costs.

In the United States, manufacturing costs typically represent only about half of total costs: the balance is accounted for by research and development, sales and marketing. These latter costs have not been included since our respondent does not impose a charge on its Mexican subsidiary for its minimal research and development activity in Mexico. Raw materials represent approximately 50 per cent of United States manufacturing costs and are only slightly less costly in Mexico since many of the raw materials required are imported. Factory overhead cost is high in the United States since the manufacturing process is highly automated, resulting in much lower overhead costs. Direct labour costs are about the same: much lower labour rates are offset by a much higher labour content because of low automation.

Now, let us take a look at the cost comparisons of autoparts. This cost comparison is for a German manufacturer of low technology light bulbs used in automobiles. The raw materials are sourced in Mexico, and are 25 per cent less costly than in Germany. The manufacturing process is fairly labour intensive and as a result the Mexican operation's lower labour costs significantly reduce the overall costs. Factory overhead is not a significant element of cost and does not vary between Germany and Mexico. As a result, overall manufacturing costs are 40 per cent lower in Mexico than in Germany. The head of one leading multinational recently said, "For every dollar you pay a Mexican engineer, you get more from him or her than you get in other societies around the world." I mention this because it is important you recognize that you can also hire good supervisory and middle-management personnel at competitive rates; however, in addition to this economic advantage, you will also obtain capable and loyal people.

Finally, we will take a look at the cost comparisons of apparel. The Mexican example is a *maquiladora* or "in-bond" operation whose output is both exported and sold in Mexico. Manufacturing costs are 25 per cent lower in Mexico. Raw material costs are about the same since the company buys globally, and has negotiated a global, delivered price. Labour and overhead costs are just over half the United States level. Direct labour rates is approximately one-eighth of the United States rate, but this is somewhat offset by lower productivity and high fringes. SG & A and distribution costs are approximately 10 per cent lower in Mexico than in the United States.

The Mexican labour force is young, adaptable and trainable. When combined with skilful entrepreneurship, this results in growing productivity both in labour and capital.

Capital productivity compares favourably with other developing countries. In Mexico one dollar invested in 1991 resulted in six dollars of value added, whereas in other economies it was lower.

For the first time, institutional investors are considering Mexico a secure place to invest long-term funds. The stability of its currency and its macroeconomic framework attracted fund managers and other investors to look for long-term capital gains by investing in Mexico.

Many companies have discovered Mexico's attractive labour costs. They have also discovered that the quality of products manufactured in Mexico is world-class. Two examples include Ford and Sony in Mexico.

A recent study by the Massachusetts Institute of Technology on automobile manufacturing plants around the world rated Ford's Hermosillo plant in Mexico number one for quality. This plant makes Mercury Tracers and Ford Escort Sedans for both the United States and Mexican markets.

North American Free Trade Agreement

The North American Free Trade Agreement (NAFTA), ratified on January 1, 1994, among Canada, the United States and Mexico, responds to a natural and logical agreement amongst neighbours. It does not imply the consolidation of commercial blocs, but rather the conformation of a negotiated agreement for regional trade liberalization. NAFTA will do away with restrictive trade practices such as quotas and trade barriers such as tariffs in progressive steps to spur trade growth amongst member countries, and consequently, with the rest of the world. The multi-dimensional impact of NAFTA on international commerce is tremendous. It is important for all manufacturers in North America to think strategically about the world's largest free trade area with a population of 370 million people, a combined total economic output or GNP of US$6.5 trillion annually and a total trade of US$225 billion. Every business person needs to consider how he can leverage his current position in this new market: how can the comparative advantage of one's size and resources be utilized to defend one's position, expand one's market share and extend one's competitive stretch into global markets.

General Electric is an interesting case study as to how one major global company is realigning itself to serve the realities of the new North American free trade area. General Electric is locating its plants strategically to serve North America. The company is moving production between its plants so that Canada focuses on specialized, flexible product lines, the United States on large automated production facilities to serve mass markets, and Mexico on product lines requiring custom attention and, therefore, medium scales of operation.

NAFTA and the current trend towards a freer and more open market in North America require some obvious strategic responses. These include the rationalization of factories and distribution centres to take advantage of more open borders and eventually of the single market. Therefore, there is a need to evaluate the location, size and scale of manufacturing plants in order to take full advantage of the single market, the evolving infrastructure, the product demands of the new regions, and finally the moves of competitors and new foreign investors. Investors should also identify opportunities for strategic alliances, mergers and acquisitions in Mexico.

These are some of the key things that a business person today need to consider in evaluating opportunities for their company in the North American free trade area.

What can other countries do to tap and exploit the huge market that NAFTA represents? They can use Mexico as an investment gateway to NAFTA member countries and other Latin American countries. Mexico, at an antipodal location, is strategically and centrally located between the United States and Canada to the North and the fast-emerging Latin American economies to the South. Potential NAFTA members include Chile, Argentina, Venezuela and Colombia.

I would like to highlight a few factors about various industries which have been detected as of main interest to foreign investors. The basis for the above is a study of the Mexican Investment Board conducted by A.T. Kearney. And as you may have gathered from the perspective I have presented to you, the philosophy of the Mexican government is that businessmen should conduct business. The government's only responsibility is to create an environment conducive to the development of profitable operations, provided that these help increase the national wealth of the country as a whole.

Market niches exist in almost all economic activities and an increasing number of business people are discovering them. Most of the above is related to the three fundamentals of Mexico's new economic platform: certainty, sustained growth and a productive working environment. However, let me highlight the industries in which international entrepreneurs have found a comparative advantage by combining the above with their technology and expertise:

1. **Apparel**
   Labour availability and competitive wage rates make the apparel industry attractive to foreign investors. Moreover, Mexico's geographical position is a plus in serving global markets.
2. **Automotive Parts and Service**
   The strong presence of big automakers in Mexico and their strategies to increase the amount of regional components produces investment opportunities in this sector. Furthermore, the replacement needs of vehicles due to the advancing age of existing cars on the road in Mexico provides for an attractive domestic market.
3. **Electronics and Communications Equipment**
   This sector is in a situation similar to the automotive parts and service industry, since increased domestic demand for more sophisticated products and the need for better means of communication create advantages for local production.
4. **Building Materials**
   The demand derived from needed infrastructure projects, as well as ecology-related ones and housing, etc., combined with competitive costs of raw materials (cement, glass, mining ores, etc.) create good investment opportunities in this area.

5. **Food Processing**
   The potential for this industry is enormous. It is favourably affected by low production costs, the availability of raw materials and an increasing demand for more sophisticated food products.
6. **Electrical Products**
   Domestic demand for household goods and appliances is expected to grow rapidly, considering a sustained increase in real income and purchasing power. Mexico's position as an export base to the rest of the Latin American countries is another important factor.

Mexican Investment Board

The Mexican Investment Board was created to assist foreign investors in exploring and pursuing business opportunities in Mexico. It is a non-profit organization jointly owned by the Mexican government and the private financial community, principally Mexico's leading banks and brokerage houses. It provides information on such matters as the different geographic regions, transportation and communications, labour availability and skills, government incentives and policies, markets and investment regulation and approvals. It also assist investors in finding the right professional services, in identifying potential joint-venture partners and suppliers, in arranging visits to Mexico, in suggesting key firms and business people in Mexico that foreign investors need to meet, and finally in dealing with the few but necessary government authorities.

11

Energy, Plastics, Chemicals and the Environment

ERIC B. FABLE

AS an active participant in the environmental industry for the last ten years, predominantly in the United States, I feel a little like the fox guarding the chicken coop . . . in that the energy, plastics and chemical industries are ongoing targets for environmental preventive and remedial activities. With that certain bias noted, however, these four industries will, without question, play a major role in the unfolding of the economic story that is taking place throughout most of the entire Pacific Rim.

The last thirty days has seen an international trade mecca of activity on a very visible plateau with NAFTA being approved by the United States Congress and President Bill Clinton almost immediately after that political success leaving for Seattle, Washington in November 1993 to host the Asia-Pacific Economic Cooperation (APEC) to discuss the economic future of the entire Pacific Rim. We, the American public, were subjected to such pronouncements as "NAFTA, the biggest and most powerful economic unit in the world" one day, and two days later the special cover report of *International Business Week* features the statement, "Asia's wealth is creating a massive shift in global economic power," and, "in ten years you will be looking at some of the truly gigantic companies of the world right here in East Asia."

For this type of cataclysmic change to occur there must be a successful integration and co-ordination of a multitude of activities, disciplines, technologies and capabilities. The sourcing of these will be global in every respect.

Concurrent with the capital required to construct the needed multinational infrastructures, there must be, for example, environmental impact analyses so as to allow these nations to avoid the kinds of pitfalls encountered by the United States in the past. The chemical, plastics and energy production industries need to work hand-in-hand with the environmental industry, on a *proactive* basis, in the new construction of operating assets as well as in the cleanup or remedy of the results of uncontrolled practices over the past years or decades. With the scale of development throughout the Pacific Rim, the opportunities for technology transfer, technology and capability teaming or collaboration, and technology sourcing are unmatched in recent history.

In the environmental industry alone, the kinetic level of research and development within the specific field of contaminated soil and water cleanup is producing more efficient and economically attractive methodologies on nearly a quarterly basis. It is a full-time activity to do nothing but stay abreast of the technological advances being announced in the biweekly periodicals. An often-utilized term in our industry is "Best Developed Available Technology (BDAT)". This is interpreted as a proven and developed remedial technology that works and provides solutions in the field, not just the laboratory, and one that has economic viability when compared with alternative methodologies. The environmental remedial BDATs that are emanating from the United States, Canada and a few countries in Western Europe is astounding. The technologies to cleanup most contaminants, both hazardous and non-hazardous, are in existence. All we are doing now is to make these processes more complete, more portable, faster and less expensive and improve the basic knowledge in existence and come up with new processes.

There are many entities in the United States looking for teaming-ups or joint-venture partnerships throughout the international arena. The American Graduate School of International Management (Thunderbird), with its main campus in Phoenix, Arizona and burgeoning foreign campuses in Europe, Asia and Latin America, through its articulate President, Dr. Roy Herberger, is promulgating and incubating technological joint ventures throughout the world. The benefits, interests and goals are mutual.

It has often been said by noted historians of the last century that if we were to learn from history mankind could save an infinite number of lives and a staggering amount of time and money. With the knowledge base, real-time work experience and desire for partnering that is available to perhaps the fastest growing area of the world, it would indeed be a black mark if it were not utilized and capitalized upon.

Whether it is the construction of a plastics manufacturing facility, a chemical processing plant or a cogeneration solid waste producer of electricity, or the cleanup of a fifty-year-old petroleum refinery site that has contaminated the surrounding soils and underlying ground water, the most current, sophisticated and cost-saving methodology is available. Hundreds of millions of dollars and tens of millions of man-hours have been expended to get to the "Best Developed Available Technology", and it would be an incredible waste not to benefit from that expenditure.

A few specific examples of existing technological capabilities within the environmental remediation area would necessarily include the rapid and ongoing replacement of incineration and landfill deposits with soil washing and biodegradation treatment trains that work in concert to truly eliminate many contaminants, particularly within the realm of petroleum products and selected organic solvents. It is projected by the United States' Environmental Protection Agency that bioremediation, as a total solution to a broad range of hazardous and non-hazardous environmental contamination sites, will be the dominant technology by the end of the decade, based upon its capability for complete elimination of the contaminant or contaminants as well as the accompanying comparative economic viability. The era of digging up contaminated soil and trucking them to a landfill or of placing endless numbers of steel drums filled with the most toxic of substances in a landfill----only to wait for the day they leak into the underlying strata and causing the expenditure of hundreds of millions of dollars to clean up—is fortunately drawing to a close.

Yes, environmental remediation is a regulatory-driven industry because the facts of life are that altruism with private industry capital is not a primary mover anywhere in the world. Fortunately, a great deal has transpired for the good since the 1989 Exxon Valdez oil-spill incident in Prince William Sound in Alaska. The lessons learned from that exasperating experience have been numerous and significant. The bastion of environmental regulations that has been created by the United States' Environment Protection Agency, although certainly not optimal, does provide a model for a significant starting point for other nations. Such is occurring throughout the world in varying degrees, and the private sector is realizing that it must contribute to both the cleanup of past pollutions and the prevention of it in the future.

The Rio Earth Summit (formerly known as the United Nations Conference on Environment and Development) in 1992, which brought together governments and representatives of non-government organizations, was an event of substantial proportions, where a global commitment was made to address such issues as biodiversity, sustainable development and climate change, but in many respects it raised more questions than it provided answers. The postulation of these questions within the context of an international forum was extremely positive, and has brought about an increased level of national environmental awareness among Eastern European countries, where pollution has reached catastrophic proportions over the last forty years, and Asean.

12

Electronics and Software Engineering

KENNETH GOLDSTEIN

STRATEGIC alliances can help not only large multinational firms, but also small- and medium-sized industries (SMIs) that wish to take advantage of the Pacific Rim's dynamic growth and investment potential. With nearly two billion potential customers, 70 per cent of the world's wealth and a US$3-billion market, the Asia-Pacific region offers tremendous opportunities for profits.

Let us discuss the critical current technology issues that will impact us well into the twenty-first century. There are three technology areas that will shape our high-tech markets, and, incidentally, our world: high-speed telecommunications, re-engineering and visualization. What do these areas have in common? Each depends on both socio-economic and technological policies and trends. And each area is directly co-dependent on local, national and international interactions, as all of them are based on the sharing of information and ideas among people.

As our ever-growing interconnected technologies "light up" our global consciousness, we will have to consider our corporate goals and objectives in new ways. High-tech industries generate positive external economies; thus, social return on resources placed in these sectors exceeds private return.

The common purpose for membership in any social, political or technological bloc has always included economic co-operation and development. But the essential element for each member country has been the issue of national interest, which embraces concepts of national identity and national economic strategy.

The United States is actively seeking global business partnerships with high-tech firms worldwide. The recent passage of the North American Free Trade Agreement (NAFTA) and the G-7 conference illustrate President Bill Clinton's desire to dissolve barriers to jobs, money, information and technology worldwide.

During the Cold War period, many countries perceived external threats that essentially superseded individual national issues. Now, as external threats are seen to diminish, the members of the "World Market" have a choice: they can limit themselves to the political management of old tensions, or they can promote freer trade and investment within their regions through economic co-operation.

It is my contention that we, as technology leaders, must closely consider the global effects of our local technology activities. Neither pure technology nor traditional business thinking will produce winning strategies in the twenty-first century. National issues such as tariff and non-tariff barriers, regional standards agreements and domestic market liberalization must be faced head-on and resolved in manners to the benefit of the entire planet.

This is not a simple thing to do from our usual two-dimensional perspective. All of us are used to looking at our high-tech markets from a very two-dimensional viewpoint. On the technology side, we ask ourselves, "How can we make this microprocessor faster, add more transistors, and eliminate bottlenecks." On the commercial side, we ask, "How can we be more globally competitive, what products can we spin off, and what kinds of new products can we develop." There are other issues to be considered here.

As I was growing up in the 1950s, I recall reading with much pleasure a book entitled *Flatland*, written in the late 1800s by a mathematician name Edwin Abbott. This book is often used by modern educators to illustrate the difficulties encountered in trying to visualize Albert Einstein's 4D space-time. The first sentence begins, "I call our world Flatland, not because we call it so, but to make its nature clear to you, my happy readers, who are privileged to live in Space."

This tragic tale is about a Square who lives in Flatland, and is visited by a Sphere. The Square is, of course, baffled by the Sphere who appears to be a circle with a constantly-changing circumference. The Sphere tries to explain three-dimensional space to the Square, who refuses to believe in any such notion until the Sphere rather disdainfully describes everything in the rooms and locked drawers of the Square's two-dimensional house. This feat is possible only by peering down from above Flatland.

Finally, the Sphere yanks the Square out of his plane for a look around. Upon his return, the amazed Square tries to explain the third dimension to his fellow polygons, but is jailed as a dangerous lunatic. While in prison, he gradually loses the ability to visualize the three-dimensional objects he saw.

We, my friends, are in a similar viewing mode, where we ponder the two-dimensional aspects of economic and technological success, with little heed for the sociological and cultural impacts. The very rapid product-development cycles and the specialized nature of our high-tech marketplace today make it essential for us to observe a two-dimensional playing field that is a great deal simpler from a conceptual standpoint than when seen from up close.

Few would argue the competitive advantage of obtaining accurate, timely business information. Access to technological, marketing and strategy information is a great equalizer, but it is not the access itself that is the challenge, but rather how an organization integrates the information into its current operating structure and processes. The "international information highway" promulgated by the Clinton administration is a dramatic illustration of how a national government can create regional standards that can positively impact multiple countries.

Technologies such as Asynchronous Transfer Mode (ATM) will transfer the way information is transferred over local- and wide-area networks. Integrated Services Digital Network (ISDN) will allow even the smallest end-users to implement electronic mail and videoconferencing over standard telephone lines. But all of these promising technologies require standardization of equipment and services to be applicable for international use.

In the development of international standards for public telecommunications networks, services and equipment, the International Telecommunications Union (ITU), is recognized as the pre-eminent standards-setting body. Other related international standards organizations include the International Electrotechnical Commission (IEC) and the International Standards Organization (ISO).

The ITU is a United Nations Specialized Agency with 176 current Member Administrations drawn together by treaty obligation. The binding nature of this treaty on industry is at the sovereign discretion of individual member countries.

Establishment of a regional, multi-country telecommunications industry coupled with the introduction of competition into the marketplace necessitates a new business structure for many national administrations. This new structure often involves privatisation of a previously state-owned agency. In order to present an attractive investment option when the stock of the new corporation is offered to the public, the national government must often institute significant productivity improvements to meet requirements demanded by potential buyers.

These productivity enhancements reflect not only a change in business structure, but also a change in organizational mindset, from one of civil service to one of private enterprise, competition and profitability. Many of this productivity gains can be realized through deployment of new systems resulting from re-engineering efforts within the organization. I define re-engineering here as the redesign of processes and systems to take advantage of new information technology.

In a competitive market, being first to deliver a new product or service, or being able to react first to a customer's need for an existing offering, is a tremendous competitive advantage. This is particularly true when dealing with major multinational customers used to wielding their strong buying power in all procurement activities.

One of the most dramatic impacts of telecommunications deregulation and the introduction of competition within a country is the new power that customers can then employ. The most significant of these powers is the freedom to go elsewhere for a given telecommunications product or service. New competitors entering the market often have the advantage in deploying new and improved systems as they have no embedded base of systems to "work around" or discard. They thus end up offering newer systems with higher levels of integration and functionality, which also improves the entire national infrastructure.

In the future, re-engineering can be viewed as a continuing process with increased emphasis placed on maximizing the new technological capabilities of major telecommunications networks, transmission media and switches, as well as improvements in support systems hardware, software and operating platforms.

The last technology area I will discuss, visualization, is the technology that promises the greatest opportunity for immediate payback for our high-tech industries. In the early days of engineering, many people viewed three-dimensional modelling, rendering, animation and all other forms of computerized visualization as nothing more than decoration: just another way to sell a product.

After all, engineers and architects have been doing without these "pretty pictures" for many years. Besides, in the real world of engineering, pretty pictures just draw attention away from the *real* product—the drawing. If you can't read a drawing, then you don't belong in this industry—right? Of course, this narrow view of things ignores artist renderings, scale models and the constant demands upon engineering to put more and more information into the drawings.

The value of visualization is not dependent only on the technical ability to communicate. There must also be something worth communicating. The more complex the information we need to communicate, the more valuable visualization becomes. The *value* of visualization is directly proportional to the *quality* of the information it conveys.

Visualization is not engineering. It's not science either. It simply communicates their results. An image can explain, describe or predict an experience, and must be presented and understood within some real-world context. Visualization provides a mechanism for immediately relating volumes of complex data to our own experience.

The newest—and most hyperbolic—application of visualization is "Virtual Reality". Virtual reality is a method of visualizing three-dimensional objects or entities by placing the user's viewpoint in a "virtual world" containing these objects and allowing entity manipulation via "electronic hands". It is really a rather straightforward application of several off-the-shelf technologies. With a few technicalities, virtual reality is based on five primary perceptual factors:

1. **Real-Time Interaction**
   The user should have the sense that the environment is responding to his actions and presence in real time.
2. **Immersion**
   The user should have a sense of being inside the virtual world, not simply observing it from the outside.
3. **Model Complexity**
   The user should perceive the virtual world as being as "real"as possible.

4. **Photorealism**
   The more visually realistic the virtual world, the more real the user will perceive it as being.
5. **Resolution**
   The required resolution is closely related to the degree of realism that is desired for the users.

In terms of immediate applications, there are a number of ways to use virtual reality technologies to minimize time-to-market:

1. **Constructibility Reviews**
   Drawings are difficult to visualize in three dimensions. However, as a result of the reality of three-dimensional computer images, the interpretation of design configurations is immediate and unambiguous. Furthermore, the three-dimensional animation system can work directly from design data to create a "snapshot" of the current design, making it much more current than even plastic models.
2. **Construction Planning and Sequencing**
   The understanding of spatial relationships among all design objects using virtual reality greatly simplify the task of determining construction and installation sequences.
3. **Construction Simulation**
   The ability to completely simulate the construction task to a high degree of accuracy and detail can significantly enhance the likelihood of success. Virtual reality adds an efficient planning tool, makes it possible to verify the construction plan at every step and provides a visual tool for communicating the construction and installation plan to all personnel.
4. **Schedule Animation**
   The link between three-dimensional design data and the critical-path schedule can also be linked using virtual re-

ality. Building a product or facility using different schedule assumptions allows visual comparison of results. The impact of schedule delays or changes can be quickly reviewed and communicated.

5. **Training**

   As technologies become more complex, training becomes more important. Virtual reality can be effective for training in construction, operations and maintenance activities based on the latest real-world activities.

6. **Remote Operations**

   Virtual reality opens the way for use of remote operations in hazardous or inaccessible environments, including hazardous waste cleanup, undersea and space construction, and modifications to nuclear facilities. There are also more opportunities as exploration and construction robotics technologies become more routine.

In summary, high-speed telecommunications, re-engineering, and visualization are three of the most critical technologies that will impact us well into the next century. Telecommunications will allow us to obtain real-time data that will improve out market strategies. Re-engineering will allow us to modify our old methods of using information. Visualization will increase our understanding of the information and reduce our time-to-market. Are these technology areas obvious? Certainly. Are they easy to implement? They can be----with foresight, planning and standards.

Acknowledgement

THE Asian Strategy & Leadership Institute wishes to acknowledge the contribution of the following people: Mirzan Mahathir, Chairman of the Organizing Committee, PRBCS, and Chairman of Sabit Sdn. Bhd.; Dr Rafael Boritzer, Co-Chairman of the Organizing Committee, PRBCS, and President of International Infotrade, Inc., United States; Michael Yeoh, Executive Director of ASLI and Sunway College; Dr Oded Shenkar, Professor of Asian Management at the University of Hawaii; James Khoo, Associate Director of ASLI; Valerie Koenig, International Infotrade, Inc., United States; Evelyn Chin, Conference Manager of ASLI; and Hasnan Mokhtar, Marketing Executive of ASLI. All were instrumental in organizing the symposium and in undertaking the publication of this book.

Contributors

Mirzan Mahathir is the Managing Director of Sabit Sdn. Bhd. and the Executive Chairman of the Asian Strategy & Leadership Institute in Kuala Lumpur, Malaysia.

Mahathir Mohamad is the Prime Minister of Malaysia.

Michael Yeoh is the Executive Director of the Asian Strategy & Leadership Institute in Kuala Lumpur, Malaysia.

George Yeo Yong-Boon is the Minister for Information and the Arts and the Second Minister for Foreign Affairs of Singapore.

William F. Miller is President Emeritus of SRI International and Professor of Public and Private Management at Stanford University's Graduate School of Business in the United States.

Farid Elashmawi is the President of Tech-Trans Global Success in San Jose, California. He is also the co-author of *Multicultural Management.*

Oded Shenkar is Professor of Asian Management at the University of Hawaii in the United States.

Peter Pond is the Commercial Counsellor and Senior Trade Commissioner of the Australian High Commission to Malaysia in Kuala Lumpur, Malaysia.

Peter Cook is Australia's Minister for Trade and Chairman of the Cairns Group.

John David Mulroney is the Commercial Counsellor of the Canadian High Commission to Malaysia in Kuala Lumpur.

Jose Landa is Executive Vice-President for Asia of the Mexican Investment Board in Mexico.

Eric B. Fable is the President of Technologies International, Inc. in Scottsdale, Arizona.

Kenneth Goldstein is the Chief Executive of Computer-Aided Technologies International, Inc. in the United States.

MANAGEMENT STRATEGIES FOR VISION 2020: DEVELOPING COMPETITIVE STRATEGIES FOR THE WINNING EDGE

Michael Yeoh

The global business environment is getting more competitive. In an unpredictable world, the importance of strategic management and planning can no longer be ignored. The greatest challenge facing management today is the strategic management of change. Corporate leaders are beginning to realize that generalized prescriptions for management actions no longer apply. To be successful, every company must develop its own core competencies and enhance its competitive advantage. Companies need to plan ahead and develop a sustainable competitive edge.

Leadership and strategic management will enable companies to successfully compete in the new business environment of the 1990s and beyond. But, surprisingly, strategic thinking, which is so vital for business success, is a rare commodity in many organizations.

This book is a practical guide to formulating and implementing business strategy and explains the crucial relationship between strategy, leadership, vision, mission, total quality and change management. The ideas expostulated are useful for managers who wish to improve the bottom-line of their organizations by improving productivity, quality, working environment, interpersonal relationships, work performance and cost competitiveness. By focusing on leading-edge processes, excellence and new management concepts, you are encouraged to think more strategically and, eventually, derive greater profitability for your company.

ISBN 967 978 551 3

PELANDUK PUBLICATIONS

VISION & LEADERSHIP: VALUES AND STRATEGIES TOWARDS VISION 2020

Michael Yeoh

It is a certainty your company will sink without management. Management---whether good or bad---will determine the course your business takes.

This book brings you the most effective management concepts, techniques, styles and tricks that need to be adopted to achieve Vision 2020.

The objective of this book is to stimulate strategic and creative thinking among corporate managers and executives on what needs to be done so that Malaysian organizations can compete more effectively in the 1990s and beyond. It is only by focusing on excellence, quality, creativity and greater productivity can we compete better and derive greater profitability for our companies---and the nation.

ISBN 967 978 552 1

PELANDUK PUBLICATIONS